Explainable Artificial Intelligence: An Introduction
to Interpretable Machine Learning

Uday Kamath • John Liu

Explainable Artificial Intelligence: An Introduction to Interpretable Machine Learning

 Springer

Uday Kamath
Ashburn
VA, USA

John Liu
Nashville
TN, USA

ISBN 978-3-030-83358-9   ISBN 978-3-030-83356-5   (eBook)
https://doi.org/10.1007/978-3-030-83356-5

This Springer imprint is published by the registered company Springer Nature Switzerland AG
The registered company address is: Gewerbestrasse 11, 6330 Cham, Switzerland

To my parents Krishna and Bharathi, my wife Pratibha, the kids Aaroh and Brandy, my family and friends for their support.

–Uday Kamath

To my wife Catherine and daughter Gabrielle Kaili-May and my parents for their encouragement and patience.

–John Liu

Foreword

The extraordinarily rapid integration of AI into many (if not most) aspects of life, entertainment, and business is transforming and disrupting the processes that power the flow of human experience. The pace of AI adoption everywhere is intense, accompanied by both immense benefits and immense risks. While many of the risks reside in the human, legal, social, and political dimensions of AI applications, the primary source of risk remains in the technical implementations of AI. Most importantly and significantly, the current technical risk landscape is populated by concerns around a taxonomy of AI issues: understandability, explainability, transparency, interpretability, trustworthiness, and more.

In this comprehensive book on explainable AI (XAI) by John Liu and Uday Kamath, we find a valuable and thorough handbook for the AI/ML community, including early learners, experienced practitioners, and researchers. The book covers the various dimensions of the AI technical risk taxonomy, including methods and applications. The result of the authors' extensive work is an in-depth coverage of many XAI techniques, with real-world practical examples including code, libraries, algorithms, foundational mathematics, and thorough explanations (the essence of XAI).

The XAI techniques presented in-depth here range from traditional white-box (explainable) models (e.g., regression, rule-based, graphs, network models) to advanced black-box models (e.g., neural networks). In the first case, XAI is addressed through statistical and visualization techniques, feature exploration and engineering, and exploratory data analysis. In the latter case, XAI is addressed through a remarkably powerful and rich set of methods, including feature sensitivity testing through dependence, perturbation, difference, and gradient analyses. Extra attention is given to three special cases for XAI: natural language processing (NLP), computer vision (CV), and time series.

The book aims (and succeeds) to bring the reader up to date with the most modern advances in the XAI field, while also giving significant coverage to the history and complete details of traditional methods. In all the examples, use cases, techniques, and applications, the consistent theme of the book is to provide a comprehensive overview of XAI as perhaps the most critical requirement in AI/ML both for now and in the future, from both technical and compliance perspectives. In that regard, the forward-looking discussions at the end of the book give us a glimpse of emerging XAI research areas that will advance our AI risk-compliance posture, including human-machine collaboration (as in assisted and augmented intelligence applications), causal ML, explainable security, responsible AI, and multidisciplinary research into explainable and interpretable intelligence technologies as they impact humanity.

I expect that this book will be an essential textbook, guidebook, reference book, and how-to book in XAI design discussions, operational implementations, risk and compliance monitoring applications, and essential conversations with technically informed AI end-users, stakeholders, and regulators.

Kirk Borne, Ph.D., Data Scientist, Astrophysicist, Top Influencer, and Chief
Science Officer at DataPrime.ai

Preface

Why This Book?

The field of explainable AI addresses one of the most significant shortcomings of machine learning and deep learning algorithms today: the interpretability of models. As algorithms become more powerful and make predictions with better accuracy, it becomes increasingly important to understand how and why a prediction is made. Without interpretability and explainability, it would be difficult for the users to trust the predictions of real-life applications of AI. Interpretable machine learning and explainability is of critical importance for the following reasons:

- We need interpretability to explain the model's working from both the diagnosis and debugging perspective.
- We need explanations for the end-user to explain the decisions made by the model and the rationale behind the decisions.
- Most datasets or models have been shown to have biases, and investigating these biases is imperative for model deployment. Explainability is one way of uncovering these biases in the model.
- Many industries such as finance and healthcare have legal requirements on transparency, trust, explainability, and faithfulness of models, thus making interpretability of models a prerequisite. In the European Union, some interpretations of the GDPR regulations claim that AI solutions must supply explanations for their conclusions. (Other interpretations say people need only be informed that automated processes are involved in decisions that may affect them.)

There is also a constant flux of new tools that fall in various categories such as application specific toolkits, visualization frameworks, and algorithm libraries. Python is currently the lingua-franca of data scientists and researchers to perform research in the area of interpretability and explainability. There are many libraries that have evolved in Python for interpretable machine learning and explainable AI in the last few years. We found a need for a single resource that gives concrete form to traditional as well as modern techniques in explainable AI through the use of existing tools and libraries for real-world applications. The work aims to be a comprehensive "go to" resource presenting the most important methods of explainability, insights to help put the techniques to use, and real-world examples with code for a hands-on experience.

This book offers its readers a collection of techniques and case studies that should serve as an accessible introduction for those just entering the field or as a handy guide for others ready to create explainable solutions in any domain. Here are some highlights:

- Comprehensive coverage of XAI techniques as a ready reference to the architectures, algorithms with essential mathematical insights, and meaningful interpretations.
- Thorough discussion on exploratory data analysis, data visualization, feature engineering, and selection necessary from a pre-hoc perspective for wide varieties of datasets such as tabular, text, time series, and images.
- Model validation and estimation techniques, visualization of model performance and selection for classification, regression, and clustering problems are discussed with examples.
- 25+ interpretable machine learning techniques, or white-box techniques, ranging from traditional to modern, state-of-the-art.
- 20+ post-hoc techniques, covering diverse areas such as visualization, feature importance, and example-based explanation techniques.
- 20+ explainable deep learning techniques which can be used in a generic or architecture-specific way for model diagnosis and explanations.
- XAI techniques from traditional to advanced in different areas such as time series forecasting, natural language processing, and computer vision.
- 20+ XAI tools and Python-based libraries bundled together in the context of real-world case studies and Google Colaboratory-based notebooks for the practitioners to get hands-on experience. The book's primary purpose is to be the single resource that addresses the theory and the practical aspects of interpretability and explainability using the case studies with code, experiments, and analysis to support.

Who This Book Is For

This book is an ideal text for AI practitioners wishing to learn methods that enable them to interpret and explain model predictions. This book addresses a large market as the interpretability problem is significant in healthcare, finance, and many other industries. Current AI/ML researchers will also find this book helpful as they integrate explainability into their research and innovation.

As it is infeasible to cover every topic fully in a single book, we present the key concepts regarding explainable AI. In particular, we focus on the overlap of these areas, leveraging different frameworks and libraries to explore modern research and the application. This book is written to introduce interpretability and explainable techniques with an emphasis on application and practical experience.

What This Book Covers

This book takes an in-depth approach to present the fundamentals of explainable AI through mathematical theory and practical use cases. The content is split into four parts: (1) pre-hoc techniques, (2) intrinsic and interpretable methods, (3) model-agnostic methods, and (4) explainable deep learning methods. Finally, a chapter is dedicated for the survey of interpretable and explainable methods applied to time series, natural language processing, and computer vision.

A brief description of each chapter is given below.

- In the **Introduction to Interpretability and Explainability** Chap. 1, we introduce the readers to the field of explainable AI by presenting a brief history, its goals, societal impact, types of explanations, and taxonomy. We provide the readers with different resources ranging from books to courses to aid the readers in their practical journey.

- **Pre-model Interpretability and Explainability** Chap. 2 focuses on how to summarize, visualize, and explore the data using graphical and non-graphical techniques as well as provide insights into feature engineering. Since time series, natural language processing, and computer vision need special handling regarding data analysis, each of these topics is covered further in detail.

- **Model Visualization Techniques and Traditional Interpretable Algorithms** Chap. 3 is a refresher of basic theories and practical concepts that are important in model validation, evaluation, and performance visualization for both supervised and unsupervised learning. Traditional interpretable algorithms such as linear regression, logistic regression, generalized linear models, generalized additive models, Bayesian techniques, decision trees, and rule inductions are discussed from an interpretability perspective with examples.

- **Model Interpretability: Advances in Interpretable Machine Learning Algorithms** Chap. 4 guides the readers to the latest advances made in the area of interpretable algorithms in the last few years overcoming various computational challenges while retaining the advantage of being transparent. The chapter covers most of the glass-box-based methods, decision tree-based techniques, rule-based algorithms, and risk-scoring systems successfully adopted in research and real-world scenarios.

- **Post-hoc Interpretability and Explanations** Chap. 5 covers a vast collection of explainable methods created to specifically address the black-box model problem. The chapter organizes the post-hoc methods into three categories: visual explanations-based, feature-importance-based, and examples-based techniques. Each technique in the category is not only summarized but also implemented on real-world dataset to give a practical view of the explanations.

- **Explainable Deep Learning** Chap. 6 presents a collection of explanation approaches that are specifically developed for neural networks by leveraging architecture or learning method from the perspective of model validation, debugging, and exploration. Various intrinsic, perturbation, and gradient-based methods are covered in-depth with real examples and visualizations.
- **Explainability in Time Series Forecasting, Natural Language Processing and Computer Vision** Chap. 7 discusses everything from traditional to modern techniques and various advances in the respective domains in terms of interpretability and explainability. In addition, each topic presents a case study to compare, contrast, and explore from the point-of-view of a real-world practitioner.
- **XAI: Challenges and Future** Chap. 8 highlights the paramount importance of formalizing, quantifying, measuring, and comparing different explanation techniques as well as some of the recent work in the area. Finally, we present some essential topics that need to be addressed and directions that will change XAI in the immediate future.

Next, we want to list topics we will not cover in this book. The book does not cover topics related to ethics, bias, and fairness and their relationships with XAI from a data and modeling perspective. XAI can both be hacked and also be used for hacking. XAI and its implications to security are not covered in this work. Causal interpretability is gaining popularity among researchers and practitioners to address the "why" part of decisions. Since this is a relatively new and evolving area, we have not covered causal machine learning from an interpretability viewpoint.

Ashburn, VA, USA   Uday Kamath

Nashville, TN, USA   John Liu

Acknowledgments

The construction of this book would not have been possible without the tremendous efforts of many people. Firstly, we want to thank Springer, especially our editor, **Paul Drougas**, and coordinator **Shina Harshvardhan**, for working very closely with us and see this to fruition. We would specifically like to first thank (alphabetical order) **Gabrielle Kaili-May Liu** (Junior, MIT, Cambridge), **Mitch Naylor** (Senior Data Scientist, Smarsh, Nashville), and **Vedant Vajre** (Senior, Stone Bridge High School, Ashburn) for their help in explainable AI libraries integration and validating experiments for many chapters described in the book. We would also like to thank (alphabetical order) **Krishna Choppella** (Solutions Architect, BAE Systems, Toronto), **Bruce Glassford** (Sr. Data Scientist, Smarsh, New York), **Kevin Keenan** (Sr.Director, Smarsh, Belfast), **Joe Porter** (Researcher, Nashville), and **Prerna Subramanian** (PhD Scholar, Queens University, Canada) for providing support and expertise in case studies, chapter reviews, and content feedback. We would also like to thank **Dr. Kirk Borne**, **Anusha Dandapani**, and **Dr. Andrey Sharapov** for graciously accepting our proposal to formally review the book and provide their perspectives as a foreword and reviews.

Contents

Notation[1]

Calculus

$\approx$	Approximately equal to
$\lvert \mathbf{A} \rvert$	L_1 norm of matrix $\mathbf{A}$
$\lVert \mathbf{A} \rVert$	L_2 norm of matrix $\mathbf{A}$
$\frac{da}{db}$	Derivative of a with respect to b
$\frac{\partial a}{\partial b}$	Partial derivative of a with respect to b
$\nabla_x Y$	Gradient of Y with respect to x
$\nabla_{\mathbf{X}} Y$	Matrix of derivatives of Y with respect to $\mathbf{X}$

Datasets

$\mathscr{D}$	Dataset, a set of examples and corresponding targets, $\{(\mathbf{x}_1, y_1), (\mathbf{x}_2, y_2), \ldots, (\mathbf{x}_n, y_n)\}$
$\mathscr{X}$	Space of all possible inputs
$\mathscr{Y}$	Space of all possible outputs
y_i	Target label for example i
$\hat{y}_i$	Predicted label for example i
$\mathscr{L}$	Log-likelihood loss
Ω	Learned parameters

---

[1] Most of the chapters unless and otherwise specified assumes the notations given above.

Functions

$f : \mathbb{A} \to \mathbb{B}$	A function f that maps a value in the set $\mathbb{A}$ to set $\mathbb{B}$
$f(\mathbf{x}; \boldsymbol{\theta})$	A function of $\mathbf{x}$ parameterized by $\boldsymbol{\theta}$. This is frequently reduced to $f(\mathbf{x})$ for notational clarity.
$\log x$	Natural log of x
$\sigma(a)$	Logistic sigmoid, $\frac{1}{1+\exp{-a}}$
$[\![a \neq b]\!]$	A function that yields a 1 if the condition contained is true, otherwise it yields 0
$\arg\min_x f(x)$	Set of arguments that minimize $f(x)$, $\arg\min_x f(x) = \{x \mid f(x) = \min_{x'} f(x')\}$
$\arg\max_x f(x)$	Set of arguments that maximize $f(x)$, $\arg\max_x f(x) = \{x \mid f(x) = \max_{x'} f(x')\}$

Variables

a	Scalar value (integer or real)
$\begin{bmatrix} a_1 \\ \vdots \\ a_n \end{bmatrix}$	Vector containing elements a_1 to a_n
$\begin{bmatrix} a_{1,1} & \cdots & a_{1,n} \\ \vdots & \ddots & \vdots \\ a_{m,1} & \cdots & a_{m,n} \end{bmatrix}$	A matrix with m rows and n columns
$A_{i,j}$	Value of matrix $\mathbf{A}$ at row i and column j
$\mathbf{a}$	Vector (dimensions implied by context)
$\mathbf{A}$	Matrix (dimensions implied by context)
$\mathbf{A}^\mathsf{T}$	Transpose of matrix $\mathbf{A}$
$\mathbf{A}^{-1}$	Inverse of matrix $\mathbf{A}$
$\mathbf{I}$	Identity matrix (dimensionality implied by context)
$\mathbf{A} \cdot \mathbf{B}$	Dot product of matrices $\mathbf{A}$ and $\mathbf{B}$
$\mathbf{A} \times \mathbf{B}$	Cross product of matrices $\mathbf{A}$ and $\mathbf{B}$
$\mathbf{A} \circ \mathbf{B}$	Element-wise (Hadamard) product
$\mathbf{A} \otimes \mathbf{B}$	Kronecker product of matrices $\mathbf{A}$ and $\mathbf{B}$
$\mathbf{a}; \mathbf{b}$	Concatenation of vectors $\mathbf{a}$ and $\mathbf{b}$

Probability

$\mathbb{E}$	Expected value
$P(A)$	Probability of event A
$X \sim \mathcal{N}(\mu, \sigma^2)$	Random variable X sampled from a Gaussian (Normal) distribution with μ mean and σ^2 variance.

Sets

$\mathbb{A}$	A set
$\mathbb{R}$	Set of real numbers
$\mathbb{C}$	Set of complex numbers
$\emptyset$	Empty set
$\{a, b\}$	Set containing the elements a and b.
$\{1, 2, \ldots n\}$	Set containing all integers from 1 to n
$\{a_1, a_2, \ldots a_n\}$	Set containing n elements
$a \in \mathbb{A}$	Value a is a member of the set $\mathbb{A}$
$[a, b]$	Set of real values from a to b, including a and b
$[a, b)$	Set of real values from a to b, including a but excluding b
$a_{1:m}$	Set of elements $\{a_1, a_2, \ldots, a_m\}$ (used for notational convenience)

Most of the chapters unless and otherwise specified assumes the notations given above.

Chapter 1
Introduction to Interpretability and Explainability

In recent years, we have seen gains in adoption of machine learning and artificial intelligence applications. However, continued adoption is being constrained by several limitations. The field of Explainable AI addresses one of the largest shortcomings of machine learning and deep learning algorithms today: the interpretability and explainability of models. As algorithms become more powerful and are better able to predict with better accuracy, it becomes increasingly important to understand how and why a prediction is made. Without interpretability and explainability, it would be difficult for us to trust the predictions of real-life applications of AI. Human-understandable explanations will encourage trust and continued adoption of machine learning systems as well as increasing system safety. As an emerging field, explainable AI will be vital for researchers and practitioners in the coming years.

This book takes an in-depth approach to presenting the fundamentals of explainable AI through mathematical theory and practical use cases. The content is split into four parts: pre-model methods, intrinsic methods, post-hoc methods, and deep-learning methods. The first part introduces pre-model techniques for Explainable AI (XAI). Part Two presents classical and modern intrinsic model interpretability methods, while Part Three details the collection of post-hoc methods. Part Four dives into methods tailored specifically for deep learning models. All concepts are presented with numerous examples to build practical knowledge. This book makes an assumption that readers have some background in elementary machine learning and deep learning models. Knowledge of the python programming language and its associated packages is helpful, but not a requirement.

© The Author(s), under exclusive license to Springer Nature Switzerland AG 2021
U. Kamath, J. Liu, *Explainable Artificial Intelligence: An Introduction to Interpretable Machine Learning*, https://doi.org/10.1007/978-3-030-83356-5_1

1.1 Black-Box problem

Innovation in machine learning algorithms has led to great advances in prediction power and accuracy. However, they have increasingly become more complex. This is an unfortunate trade-off between improved quality and transparency. We may be able to observe the set of outputs for a given set of inputs to a model, without knowledge or understanding of its internal workings. Unlike mathematical models that have inherent structure, machine learning models can learn the mapping of inputs to outputs directly from the data. For some models like decision trees, this mapping is easily discernible. For others like random forests or deep learning models, it becomes next to impossible to understand how predictions are made. Many machine learning and deep learning models are essentially "black-boxes" that do not reveal the internal mechanisms and nuances to their predictions (Fig. 1.1).

Fig. 1.1 Black-Box
algorithm lacks transparency

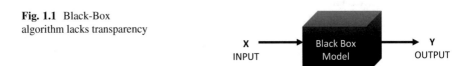

This lack of transparency and understanding can have serious consequences to our trust and adoption of these models. For instance, how do we know if the model predictions may be wrong? This is especially important in high-stakes domains such as healthcare. Would a doctor or patient trust a cancer prediction if a trained model has an accuracy of 99 percent? What if, unknown to us, the model misses the most-malignant cases? What if the high accuracy was due to data-leakage in the test data, such that out-of-sample performance was much worse? This is why explainable AI is a vital to our adoption of machine learning. For high-stakes decisions such as credit loans, discriminating bail and parole applications, medical diagnosis, etc., it becomes imperative for the machine learning models to be explainable [Kle+18, Lak+19].

1.2 Goals

Explainable AI (XAI) seeks provide us insight on the decision-making ability of an AI system. It helps us to understand how, when, and why predictions are made. Consequently, it can build greater trust and improve the safety of our use of AI models, encouraging their greater adoption in our society. We begin our exploration of XAI by defining several inter-related goals: understandability, comprehensibility, interpretability, and transparency. Each of these concepts is closely tied to model complexity. While many of these may vary or overlap across different domains, they are distinct in their desired outcomes, characteristics, and/or approaches (Fig. 1.2).

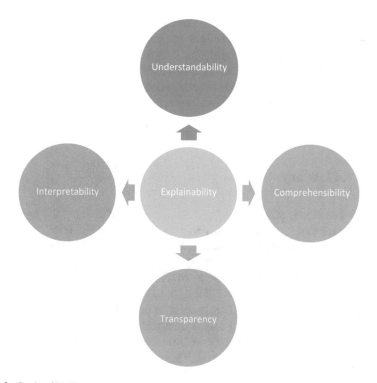

Fig. 1.2 Goals of XAI

1. **Understandability**: Understandability is the notion that, to be useful, the underlying function of an AI model must be understandable to humans. The concept of understandability, also known as intelligibility, is the property of the overall model to be understandable without any need for details and explanation of its internal algorithmic structure used by the model. For instance, the function of autoencoders is easy to understand, even without intimate knowledge of how autoencoders compress and uncompress inputs [RHW86].
2. **Comprehensibility**: Pertaining to ML models, comprehensibility refers to the ability of a model to represent and convey its learned knowledge in a human-understandable fashion. In general, measuring how well humans can understand explanations is difficult in a nominal sense, but somewhat easier from a relative perspective. For instance, it is hard to quantify how much the principal components derived in PCA are human understandable, but we can likely say factors derived in factor analysis are generally more comprehensible [Shl14].
3. **Interpretability**: Interpretability, often used interchangeably with explainability, is the ability to explain or provide meaning to model predictions. In particular, the goal of interpretability is to describe the structure of a model in a fashion easily understandable by humans. That is, for a model to be interpretable, it must

be describable in simple terms for a human to understand. As interpretability is a subjective notion, it often depends on the audience and context.

4. **Transparency**: A model is transparent if its internal structure (structural transparency) and algorithm (algorithmic transparency) by which it makes predictions is understandable. Transparency helps us comprehend the basis of a model and addresses the question of why a model works the way it does. It is worth noting that a model can have different degrees of understandability.

1.3 Brief History

The field of machine learning modeling has evolved rapidly over the past century. Many computational models were created to model real-life biological and cognitive processes, and the advent of the computer launched an explosion of new algorithms that previously were constrained by computation power. This trend continues to today, with the increasing adoption of High-Performance-Computing (HPC) clusters that can perform at 4 peta-FLOPS, or 4,000,000,000,000,000 floating point operations per second (for reference, there are only about 86,000,000,000 neurons in the human brain) [Zha19]. Our notion of machine learning has evolved over the past few decades as computation power increased, from the early expert systems to the current deep learning algorithms. This evolution generally achieved greater accuracy at the expense of complexity and explainability.

1.3.1 Porphyrian Tree

Explainable models have existed for a long time before the modern invention of the computer with its data processing capability. One of the earliest examples is the decision tree, a prediction and classification algorithm with intrinsic explainability. The decision tree algorithm is based on the notion of recursively partitioning data using their characteristics to segregate into groups with similar target values (Fig. 1.3).

Perhaps the earliest documented implementation of the decision tree is attributed to Porphyry of Tyre, an influential Phoenician neoplatonic philosopher known for his work "Introduction to Categories" which incorporated Aristotle's logic into Neoplatonism [Bar03]. The Porphyrian tree, as shown in the figure below, was created by Porphyry as a visual means to classify genera into species [Dar17].

As the figure illustrates, the intrinsic interpretability of decision tree predictions is readily evident in its visual, hierarchical structure. More recently, the decision tree model was alluded to by Fisher in 1936 [Fis36] and characterized by Belson in 1959 [Bel59]. It was not until 1963 and 1972 that the first regression tree was invented by Morgan and Sonquist and the classification tree was invented by Messenger and Mandell, respectively [MS63, MM72].

Fig. 1.3 Tree of Porphyry

1.3.2   Expert Systems

Beginning in the 1970s, computer scientists sought to develop models that could emulate the decision-making of human experts in a variety of fields. These expert systems were designed to be able to solve complex problems using logic and reasoning. An important consideration of these systems was that decisions were explainable, as the rules that defined the expert system were intuitive and could be easily understood (Fig. 1.4).

Unfortunately, expert systems had significant limitations in what they could achieve. Among other things, they were slow, dificult to train, and unable to deal with in changing environments. These limitations led them to fall out of favor in the late 1980s and precipitate a period known as the second AI winter [Nil09].

1.3.3   Case-Based Reasoning

As interest in expert systems declined, attention turned toward case-based reasoning models that could solve new problems by using solutions of similar problems learned in the past [WM94]. These models had a clear advantage in that their decisions were implicitly explainable as well as generalizable beyond previously seen

Fig. 1.4 XCON expert system

data. Their main criticism is that there are no guarantees that such generalizations are correct if data is scarce or imbalanced.

1.3.4 Bayesian Networks

In 1985, a new approach to probabilistic reasoning was presented by Judea Pearl [Pea85]. He presented Bayesian networks as a type of probabilistic graphical model comprised of nodes and directed edges. Bayesian network models use mathematical graphs to capture conditionally dependent and independent relationships between independent and target variables. Models can be created by experts or learned from data and then used for inference to estimate the probabilities for subsequent events. Bayesian models intrinsically have explanatory power, since they capture and are able to express the conditional relationships between variables. They have led to significant work in modeling real-world causal relationships, but popularity remains muted by the tremendous computational load needed to process large networks or datasets (Fig. 1.5).

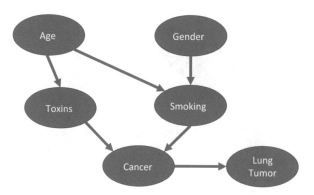

Fig. 1.5 Bayesian network example

1.3.5 Neural Networks

Alongside Bayesian networks, neural networks have taken off over the past decade with several monumental breakthroughs in deep learning and computation at scale. Neural networks can now achieve superhuman capability in many tasks in domains such as computer vision [He+15], natural language processing [Wan+20], and game-play [Mni+13]. However, deep learning algorithms tend to suffer from limited scope and it remains to be proven that they can generalize well in the real-world. Their frequent complaint and limitation are that they lack transparency and it is very difficult for practitioners to entrust them for inference ("the black-box problem").

1.4 Purpose

AI presents a number of significant issues that encompass practical, ethical, philosophical, and equitable considerations. Explainable AI methods can address and mitigate these issues in many ways, and the success of AI applications will be likely driven by explainable AI methods going forward (Fig. 1.6).

1. **Informativeness**: AI models in practice exist for the intent of augmenting decision-making in the real world. AI models are designed to achieve specific quantitative objectives, but sometimes these objectives may not match their original intent. When this happens, the consequences could be catastrophic. We rely on explainable AI to inform us of the inner relations of a model, which allow us to evaluate if or when objectives may be misaligned, misguided, or counterproductive toward our decision-making intents.
2. **Trustworthiness**: According to NIST [Phi+20], the trustworthiness of an AI application is ultimately derived by its explainability. We attribute greater trust to AI algorithms that are relevant, easy to understand, and not prone to

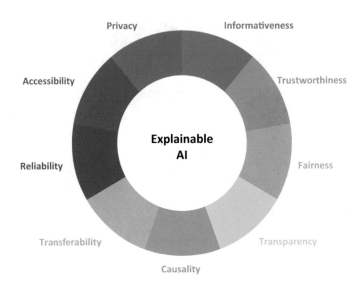

Fig. 1.6 Purpose of explainable AI

misrepresentation. An increased level of trust directly leads to better adoption
by humans. For instance, our trust with autonomous vehicles may be limited
as the methods driving the steering algorithms under the hood (literally) are not
transparent to riders. As time progresses, and we gather more information on how
autonomous vehicles behave in normal and rare situations, our level of trust will
rise in conjunction with our level of understanding of its algorithms. Lakkaraju
et al. show how user trust can be manipulated by explanations in the black-box
models by creating a framework for understanding and generating misleading
explanations that can be verified by experts [LB20].

Maister, Green, and Galford [MGG01] devised the trust equation as guiding
principle for how humans perceive trust with each other. It has application in how
we perceive trust with AI applications, such as how safe we feel when interacting
with them and whether we believe their focus is aligned with our best interests
(Fig. 1.7).

Fig. 1.7 Trust equation

$$\underset{\text{Trust}}{T} = \frac{\underset{\text{Credibility}}{C} + \underset{\text{Reliability}}{R} + \underset{\text{Intimacy}}{I}}{\underset{\text{Self-orientation}}{S}}$$

While one purpose of an explainable AI model is increased trustworthiness,
there is a trade-off between building trust and model explainability. Understanding that a model is reliable and will always act in our interests does not

automatically imply high fidelity of explanations. Trust is difficult to quantify, and we sometimes equate trust to our confidence that the model will act as intended. There is also a distinction between trusting an AI model and the trustworthiness of an AI model. The first is an expression of human attitudes, while the second is a measure of the extent to which a model can reliably serve its intended purpose. For example, we generally attribute greater trust to news stories on social media platforms than we should, even while some are untrustworthy and false. At the same time, we generally trust scientific journals far less than justified, even while their trustworthiness is high due to the peer-review process.

3. **Fairness**: Fairness is defined the impartial and just treatment or behavior in absence of any favoritism or discrimination. In the past few years, fairness in AI has come to the forefront, with important research in both data bias [Beg+20, Nto+20] and algorithmic bias [GSC18]. Our societal obligation to address fairness makes it an important goal in AI, as explainability permits us to identify if/when bias exists in the model. Explainable AI offers us the capacity to achieve and guarantee fairness in real-world AI applications (Fig. 1.8).

Fig. 1.8 Bias and explainable AI

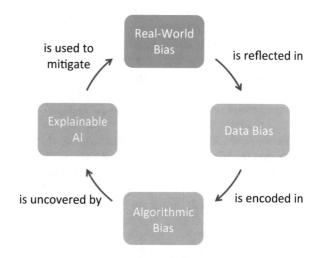

4. **Transparency**: It is often said that transparent AI is explainable AI. Model and algorithmic transparency helps us understand how particular decisions are made and is an essential part in how we build trustworthiness. However, transparency does not necessarily imply fairness or explainability. Consider an AI algorithm used to predict creditworthiness of potential borrowers. Transparency allows us to identify which features (e.g., income, education level) influence the underlying decision process, but it does mean the model is fair toward minority populations absent in the training data [Meh+19]. Nor does it actually explain why a borrower is creditworthy or not (e.g., what if they made slightly more vs having a high education degree).

5. **Causality**: One of the fundamental limitations of machine learning and AI today is the lack of causation inherent in modeling. Modeling techniques inherently leverage correlation, but ignore time or causal flow. Explainable AI is increasingly being purposed toward identifying causal relationships in the data [JMB20, Hol+19]. While significant domain and background knowledge is generally required to prove causality, explainability can be used to explore cause and effect. There is tremendous opportunity for explainable AI to tackle causal effects.

6. **Transferability**: Transfer learning is the notion that a model trained on one task can be generalized and used as a starting point for other tasks. We like to build models that are transferable since it allows us to leverage the pre-existing knowledge learned in previous tasks. Not every model is transferable, and understanding the limitations of when/how models can transfer to other tasks is an important purpose. Explainable AI allows us to understand the internal structure and learning process of a model which facilitates our ability to apply the model to other tasks. It also allows us to identify and understand what boundaries and limitations may exist in a model that affects its transferability [Rai19].

7. **Reliability**: As stated earlier, the trustworthiness of a model depends on how reliable and confident we feel in its decision-making process. Reliability and stability are desirable characteristics in an AI model so that we can expect it to make the same decision in the same circumstances. Similarly, robustness is equally desirable in our expectation for an AI model to make similar decisions in similar circumstances. Explainable AI can provide us insight into how reliable or robust a model will operate under various conditions.

8. **Accessibility**: The accessibility of AI applications by non-technical folks plays an important role in increasing popularity and adoption. Explainable AI can facilitate the knowledge and understanding of complex AI models and thereby reduce the burden by ordinary people when dealing with them [WR20].

9. **Privacy**: With privacy and security growing in importance with AI applications, one of the benefits enabled by explainable AI is the ability to assess privacy. With model explainability, we can more readily evaluate whether or not privacy is breached in encrypted representations or algorithms [VM20]. Differential privacy, another growing sub-field, seeks to maintain privacy at the origin throughout computation (e.g., adding two numbers without ever knowing what the actual numbers). Explainable AI can play an important role to ensure the integrity of differentially private models and algorithms without knowledge of the data.

1.5 Societal Impact

AI applications can have great societal impact, improving our societies and building a better world. Explainable AI can facilitate our greater adoption of AI applications by empowering us to address important issues like fairness, bias, verifiability, safety, and accountability.

1. **Fairness and Bias**: As adoption of AI models to support human life is increasing exponentially, explainable AI will be a valuable tool to uncover unfair or unethical algorithms. There are many famous cases to underscore the importance of fairness in AI systems. We have seen the deleterious effects of algorithms that exhibit gender bias [Lea18, Lea+20, FP21] and racial bias [IG20, Tho19]. COMPAS, the recidivism prediction algorithm, is a prominent example of how bias in the data was compounded by a lack of algorithmic transparency resulting in an algorithm that explicitly encoded racial and gender prejudices [RWC20, KH19]. Recently, OpenAI released the GPT-3 model, consisting of 175 billion parameters trained on Open Crawl [Bro+20] and Wikipedia. Researchers quickly observed the model exhibited serious biases, including gender, race, and religion [AFZ21, Bro+20].

    Recently, many approaches have been introduced to approach fairness in AI, including bias detection, bias mitigation, bias explainability, and simulation frameworks to understand long-term impact of algorithmic behavior [Fer+20]. With the increase in the underlying explainability of these algorithms, it becomes much easier to track down the biases and make necessary interventions to ensure fairness.

    As AI research evolves, it is becoming increasingly important to develop not just more accurate systems but also fair ones.

2. **Safety**: As we seek greater adoption of AI models, we must ensure they do not inadvertently or maliciously make decisions or take actions that are unsafe to humans. For any task, we start with a set of desired goals (e.g., shortest path traversing from here to destination) and create a system design (e.g., autonomous-driving algorithm). How do we ensure the behavior of this system design does not harm humans (e.g., strike the bicyclist in our path)? AI Safety deals with designing systems to avoid unintended and harmful behavior that may emerge from poorly designed AI systems in the real-world [JSB20, Amo+16]. A model is never completely testable in the real-world as one cannot create a complete list of scenarios in which a model might see. Explainable AI becomes a necessary prerequisite to help identify fail states in the model. For instance, it allows us to identify potential blind-spots in vision-based autonomous-driving systems, or where an AI system to predict cancer treatments may make dangerous recommendations that can harm patient health.

3. **Verifiability**: Verification is a set of powerful mathematical techniques that guarantee the correctness of an AI model, such as ensuring that certain properties are met. Importantly, it allows us to identify cases where a model may fail, or not have an explanation. Rigorous testing and training help build robust machine learning systems, but no amount of testing will formally guarantee that a system behaves as intended. In real world situations, enumerating all possible outputs for a given set of inputs is an impossible task. Verification in AI allows us to compute bounds for an AI model output that can be helpful in designing a more

resilient AI system, or a safer one [Bru+20]. Explainability is a key ingredient in verification, as it allows us to formulate verification as a computationally tractable optimization problem.

4. **Accountability**: Accountability is the ability to acknowledge and attribute responsibility for decisions and actions made by AI systems. It is an important aspect of the trustworthiness of AI models, and is closely related to transparency in these models. We may find a model to be unfair or unsafe, but we need accountability to understand why the model exhibits such behavior. Explainable AI gives us the ability to account for why individual inputs lead to such predictions, or why the overall model tends to behave in a certain way. We should note that increased transparency does not always improve accountability. Just because we have perfect clarity into the algorithm and weights of a convolutional neural network does not necessarily allow us to attribute responsibility into its behavior.

   In a broader sense, accountability in AI can serve as a tool that allows us to hold companies and organizations accountable for the performance of their AI applications in real life [Dos+19]. From the perspective of equity, AI accountability enabled by explainable AI is essential for algorithmic justice.

1.6   Types of Explanations

Explainable AI methods can provide different types of explanations to help us interpret complex systems. We list five types of explanations enabled by explainable AI to aid our understanding (Fig. 1.9).

1. **Global Explanations**: The most common question we tend to have is "how does a model work?" Global explanations serve to explain how models arrived at their predictions and can be in the form of visual charts, mathematical formulae, or model graphs. Global explanations are holistic, with the goal of providing us the ability to develop a top-down mental representation of the behavior of the model.
2. **Local Explanations**: Once we answer the question of how, we tend to ask the question why. Local explanations are bottom-up and seek to answer the question of why a model arrives at a prediction for a given input. They can attribute a prediction to specific features of the data or model algorithm.
3. **Contrastive Explanations**: Contrastive explanations help us by understanding why a model makes a certain prediction instead of another for a given input. They answer the question of "why-not" or "why X and not Y" and are often used jointly with "why" explanations to understand a model's prediction and its expected behavior. They are especially useful in determining what minimal changes in inputs or model parameters are required to cause the model to make a different prediction.
4. **What-if Explanations**: As in the classic sense, sensitivity analysis are what-if explanations of the changes in model output as we tweak inputs and model

Fig. 1.9 Types of explanations

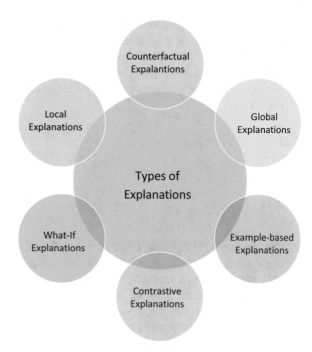

parameters. They are very useful for helping us to understand the relationships between model predictions and model features.

5. **Counterfactual Explanations**: Counterfactual explanations tell us the hypothetical changes to the input or parameters of a model that would lead the model to make a specific different input. They answer the question of "how to" arrive at a desired outcome by describing the smallest changes to the model that can be made, without needing to understand the model internal structure.

6. **Example-based Explanations**: Sometimes, it is easier to explain the behavior of a model or underlying data distribution simply by highlighting particular instances of the data. This is known as explanation by example. Common practice is to present similar input instances from which the model will predict similar outputs.

1.7   Trade-offs

According to the No-Free Lunch Theorem, every algorithm performs equally well when their performance is averaged across all possible problems. This does not mean all is lost, as knowledge of the underlying problem, data, and environment can help inform more optimal approaches. But because of the theorem, model selection will come with trade-offs.

   Similarly, while explainable AI contributes many benefits, it does not do so
without trade-offs. It is important to understand the limitations of different XAI
methods in order to recognize when one set of methods may be more relevant or
accurate over others. We discuss here the broad scope of these trade-offs, and will
delve deeper into the characteristics of individual XAI methods in later chapters
(Table 1.1).

Table 1.1 Trade-offs in
explainable AI

Property	Trade-off
Completeness	Interpretability
Efficacy	Privacy
Human explanations	Accuracy

1. **Completeness vs Interpretability**: A handful of methods such as generalized
   linear models and decision trees are inherently interpretable in that they are
   self-explanatory by construction and can provide useful explanations directly by
   inspection. However, these methods apply well to a very limited set of problems
   in the real-world. On the other hand, the Universal Approximation Theorem
   states that deep neural networks are able to approximate any continuous non-
   linear function (provided we can train them to learn the function). Unfortunately,
   these deep models are usually not transparent or easily interpretable. This
   is a common trade-off that we see with explainable AI methods—the more
   interpretable they are, the less likely they provide complete explanations of the
   AI system. Stated another way, a trade-off exists between accuracy of model
   prediction ("the what") and model interpretation ("the why"). It is hard to achieve
   both interpretability and completeness at the same time except in a handful of
   cases. The most accurate explanations are not easily interpretable by humans and
   the most interpretable explanations usually do not have complete coverage. The
   challenge in explainable AI is to generate explanations that are both complete
   and interpretable.
2. **Efficacy vs Privacy**: Increasingly, government regulatory frameworks such as
   GDPR are enforcing data privacy as an inherent consideration in real-world
   systems. This requirement for privacy can adversely limit explainability in
   these systems. The trade-off between explanation efficacy and model privacy
   is complex, as models are generally trained on a mixture of private and non-
   private data. Consider a model trained on a mix of public and private data.
   Without intervention, private data easily leak into model explanations. Adjusting
   explanations to filter out private data can be a complex task and lead to
   incomplete explanations that sacrifice accuracy.
      Recent research has aimed to reduce or eliminate this trade-off using encryp-
   tion and/or novel privacy-preserving machine learning methods. These methods
   generally come with an additional computational burden, though advances in
   computational power have and continue to mitigate this cost.

3. **Human Explanations vs Accuracy**: Even in the case where a model exhibits perfect transparency and we can easily observe the features that influence its decision-making ability, it does not mean that the model is easily understandable to humans. A trade-off exists in between the ability for a model to provide comprehensible explanations and the accuracy of the model. For instance, humans have a difficult time understanding and interpreting non-linear functions. Certain XAI techniques allow us to assume linearity for a small bounded region of a function (e.g., all continuous functions are linear if you look close enough), providing us with sensitivity analysis that is easily understood. Other XAI methods allow us to use surrogate models that can capture model behavior.

1.8 Taxonomy

Explainable AI methods has proliferated significantly in the past few years. Figure 1.10 represents a taxonomy of the family of methods based on their approach and characteristics. As new methods are being developed every day, we expect this taxonomy to increase over time.

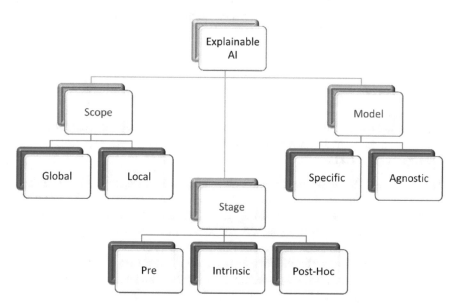

Fig. 1.10 Taxonomy of explainable AI

1.8.1 Scope

Explainable AI methods can be either global or local in scope. Some methods can extend to both. Global methods are useful what we want to interpret the macro behavior of models, whereas local methods are handy when we want to understand behavior at the micro level.

1. **Global Methods**: Global methods seek to explain the predictions of the overall model from a comprehensive, top-down approach. As a result, explanations provide an understanding of how the structures and parameters of the model lead it make predictions. This allows us to comprehend the entire model all at once by providing an understanding of how the model maps input data to features to outputs. In doing so, we gain transparency into the inner mechanisms of a black-box model.
2. **Local Methods**: Local methods, as the name implies, seek to explain how a specific sample is mapped to its output by providing us an understanding of how the model arrived at its prediction. This explains to us the rationale via the contribution of features for a specific prediction from an input, and can accomplished by approximating a model in a small region of interest using a simpler model. For instance, a local method for an image classification model can help identify the specific portions of the image that contribute to the model class prediction.

1.8.2 Stage

XAI methods can categorized based on stage—whether they are applied before, during, or after a model makes its prediction. We describe the characteristics of each below (Fig. 1.11).

1. **Pre-Model**: Pre-model interpretability techniques are independent of the model, as they are only applicable to the data itself. Data visualization is critical for pre-model interpretability, consisting of exploratory data analysis techniques.

   Pre-model interpretability usually happens before model selection, as it is also important to explore and have a good understanding of the data before thinking of the model. Meaningful intuitive features and sparsity (low number of features) are some properties that help to achieve pre-model data interpretability.

   We cover pre-model methods in Chap. 2 by delving into its relationship with EDA, feature engineering, and data/feature visualization.
2. **Intrinsic**: Intrinsic interpretability methods refer to self-explanatory models that leverage internal structure to provide natural explainability. The family of intrinsic models include basic methods such as decision trees, generalized linear, logistic, and clustering models. Natural explainability comes at a cost, however, in terms of model accuracy.

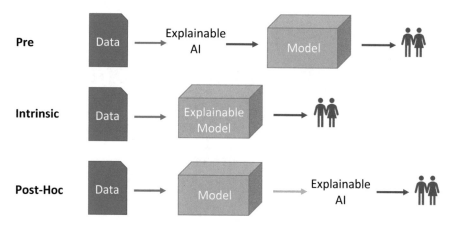

Fig. 1.11 Explainable AI categories by stage

    In Chap. 3, we cover traditional intrinsic explainability methods and investigate more advanced intrinsic methods in Chap. 4.

3. **Post-Hoc**: Post-hoc (post model) interpretability methods represent a collection of techniques that are applicable to any trained black-box models, without the need for understanding their internal structures. They provide explanations of the global or local behavior of models by resolving relationships between input samples and their predictions. Post-hoc methods are applicable even to intrinsic models.

    In Chap. 5, we discuss the wide range of post-hoc explainability methods available. We subdivide them by their approach to explanation, including visual, feature relevance, surrogate, and example-based explanations.

4. **Model Agnostic vs Specific**: Most pre- and post-hoc explainability methods are model-agnostic in that they are applicable to a wide collection of models. Some, especially with regard to deep neural networks, are model specific and apply only to a specific set of models (e.g., convolutional neural networks). Model-specific methods provide advantages over model-agnostic methods as they leverage specific characteristics or architecture of the model to provide improved explainability that may not be possible with model-agnostic methods.

    In Chap. 6, we delve in to model-agnostic and model-specific methods deep for neural networks. Finally, in Chap. 7, we examine explainable AI methods in practice and apply them to a variety of case studies in different domains.

1.9   Flowchart for Interpretable and Explainable Techniques

Figure 1.12 provides a flowchart for exploring the XAI methods discussed in this book.

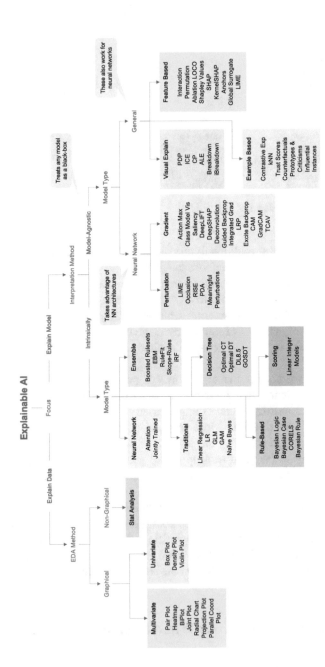

Fig. 1.12 Explainable AI flow chart

1.10   Resources for Researchers and Practitioners

There is a myriad of resources in the form of GitHub pages, survey research papers, books, and courses on the topic of XAI. Though it is difficult to list everything, we will highlight some which we have found to be very useful.

1.10.1   Books

Here we recognize various books that touch multiple areas of XAI that we think will be useful for the readers. Many of these books are available free online, and we have provided the links.

1. An Introduction to Machine Learning Interpretability by Patrick Hall and Navdeep Gill.
2. Interpretable Machine Learning by Christoph Molnar. https://christophm.github. io/interpretable-ml-book/
3. Fairness and Machine Learning by Solon Barocas, Moritz Hardt, and Arvind Narayanan. https://fairmlbook.org/
4. Explanatory Model Analysis by Przemyslaw Biecek and Tomasz Burzykowski. https://ema.drwhy.ai/
5. Responsible Machine Learning by Patrick Hall, Navdeep Gill and Benjamin. https://www.h2o.ai/resources/ebook/responsible-machine-learning/
6. Explainable AI: Interpreting, Explaining, and Visualizing Deep Learning by Wojciech Samek, Grégoire Montavon, Andrea Vedaldi, Lars Kai Hansen and Klaus-Robert Müller.
7. Explanatory Model Analysis: Explore, Explain, and Examine Predictive Models by Przemyslaw Biecek, Tomasz Burzykowski.

1.10.2   Relevant University Courses and Classes

Some relevant courses and classes with many helpful videos and lecture notes that discuss XAI topics are listed below:

1. Interpretability and Explainability in Machine Learning https://www.hbs.edu/ faculty/Pages/item.aspx?teaching=266
2. Introduction to Responsible Machine Learning https://jphall663.github.io/ GWU_rml/
3. Trustworthy Deep Learning https://berkeley-deep-learning.github.io/cs294-131-s19/
4. Data Ethics https://ethics.fast.ai/syllabus/

5. Methods of explainable AI https://human-centered.ai/methods-of-explainable-ai/
6. Interpretability and Explainability in Machine Learning https://interpretable-ml-class.github.io/
7. AI Interpretability and Fairness https://cs81si.stanford.edu/
8. Explainable AI https://www.cis.upenn.edu/~ungar/CIS700/

1.10.3 Online Resources

There are excellent online resources with a collection of articles, books, tools, datasets, etc., all assembled in one place. Some of the links are:

1. https://github.com/jphall663/awesome-machine-learning-interpretability
2. https://github.com/lopusz/awesome-interpretable-machine-learning
3. https://github.com/pbiecek/xai_resources
4. https://github.com/h2oai/mli-resources
5. https://github.com/andreysharapov/xaience

1.10.4 Survey Papers

Following is the list of survey papers which the readers can find very helpful to get an overview and the current trends,

1. Opportunities and Challenges in Explainable Artificial Intelligence (XAI): A Survey by Das and Rad [DR20].
2. Peeking Inside the Black-Box: A Survey on Explainable Artificial Intelligence (XAI) by Adadi et al. [AB18].
3. Explainable Artificial Intelligence (XAI): Concepts, taxonomies, opportunities, and challenges toward responsible AI by Arrietta et al. [Arr+20a].
4. Explainable Artificial Intelligence Approaches: A Survey by Islam et al. [Isl+21].
5. Interpretable machine learning: definitions, methods, and applications by Murdoch, W. James, et al. [Mur+19].
6. Interpretable Machine Learning—A Brief History, State of the Art and Challenges by Molnar et al. [MCB20a].

1.11 Book Layout and Details

To understand the interpretability and explainability techniques throughout the book, we have used following datasets, and here are the details.

1. **Classification**: Pima Indian Diabetes dataset is originally from the National Institute of Diabetes and Digestive and Kidney Diseases [Smi+88]. The classification dataset intends to diagnostically predict whether or not a patient has diabetes based on specific symptomatic measurements incorporated as features in the dataset. The datasets consist of several medical predictor features which are numeric such as *SkinThickness, BMI, Pregnancies, Insulin, Glucose, Age, BloodPressure, DiabetesPedigreeFunction* and one target *Outcome* classifying the patient as diabetic or non-diabetic.
2. **Regression**: The medical claims dataset created for the book—Machine Learning with R by Brett Lantz—uses demographic statistics from the US Census Bureau, reflecting real-world conditions [Lan13]. The dataset has instances of beneficiaries currently enrolled in the insurance plan with features indicating characteristics of the patient, such as *age, sex, bmi, children, smoker, Region* and the total medical expenses charged to the plan for the calendar year as the target *charges*.
3. **Time series**: Mauna Loa time series dataset has one of the longest continuous series since 1958, and measuring the mean carbon dioxide as parts per million (ppm) every month at Mauna Loa Observatory, Hawaii [Tan+09]. We use this for our univariate time series analysis through different interpretable and explainable techniques.
4. **Computer Vision**: Fashion-MNIST is a dataset of Zalando's article images, where each image is a 28×28 grayscale images, associated with a label from 10 classes—*T-shirt/top, Trouser, Pullover, Dress, Coat, Sandal, Shirt, Sneaker, Bag*, and *Ankle Boot* [XRV17].
5. **NLP and Text**: LitCovid is a curated dataset providing central access to a large number of relevant articles in PubMed that can be categorized into eight categories—*General, Forecasting, Transmission, Case Report, Mechanism, Diagnosis, Treatment*, and *Prevention* [CAL20b, CAL20a]. We will use subset of this dataset for pre-hoc exploration and post-hoc NLP-based explainability techniques.

1.11.1 Structure: Explainable Algorithm

Throughout the book we have tried to keep a consistent format for describing the pre-model, intrinsically interpretable algorithms and post-hoc explainable techniques. Each technique is described sufficiently with references and equations, plots and outputs from the algorithms when applied to the datasets, how to interpret the plots and the observations. An example with a simple linear regression model applied to the insurance dataset with just one feature is described below.

1.11.1.1   Linear Regression

Linear regression is one of the oldest techniques that predicts the target using weights on the input features learned from the training data [KK62a]. The interpretation of the model becomes straightforward as the target is a linear combination of weights on the features. Thus linear regression model can be described as a linear combination of input $\mathbf{x}$ and a weight parameter $\mathbf{w}$ (that is learned during training process). In a d-dimensional input ($\mathbf{x} = [x_1, x_2, \ldots, x_d]$), we introduce another dimension called the bias term, x_0, with value 1. Thus the input can be seen as $\mathbf{x} \in \{1\} \times \mathbb{R}^d$, and the weights to be learned are $\mathbf{w} \in \mathbb{R}^{d+1}$. The label or the output y which is a quantitative or numeric value is defined by

$$y = \sum_{i=0}^{d} w_i x_i \tag{1.1}$$

Interpreting linear regression model can be summarized as below

- Increasing the continuous feature by one unit changes the estimated outcome by its weight.
- Intercept or the constant is the output when all the continuous features are at value 0 and the categories are in the reference default (e.g., 0). Understanding intercept value becomes meaningful for interpretation when the data is scaled with mean value 0 as it represents the default weight for an instance with mean values.

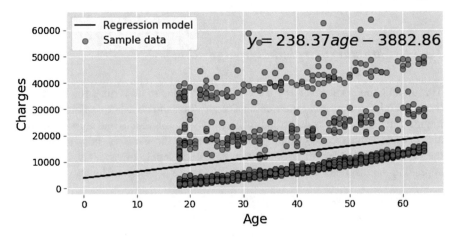

Fig. 1.13 Linear regression model with just one feature—age

Table 1.2 Explainable
Properties of Linear
Regression

Properties	Values
Local or global	Global
Linear or non-linear	Linear
Monotonic or non-monotonic	Monotonic
Feature interactions captured	No
Model complexity	Low

Observations:

- Fig. 1.13 shows that features *age* has a linear relationship with *charges* with a positive trend, i.e., as the *age* increases the *charges* increase.
- The bias or the intercept is -3882.86 while the weight for *age* feature is $+238.37$, indicating a huge positive influence of age on the insurance charges.

Explainable properties of linear regression are shown in Table 1.2.

We have made all the datasets and Python-based Google Colab notebooks available for the readers to experiment on https://github.com/SpringerXAI.

References

[AFZ21] A. Abid, M. Farooqi, J. Zou, *Persistent antiMuslim bias in large language models* (2021). arXiv:2101.05783 [cs.CL]

[AB18] A. Adadi, M. Berrada, Peeking inside the blackbox: a survey on explainable artificial intelligence (XAI). IEEE Access **6**, 52138–52160 (2018)

[Amo+16] D. Amodei et al., *Concrete problems in AI safety* (2016). arXiv:1606.06565 [cs.AI]

[Arr+20a] A.B. Arrieta et al., Explainable Artificial Intelligence (XAI): Concepts, taxonomies, opportunities and challenges toward responsible AI. Information Fusion **58**, 82–115 (2020)

[Arr+19] A.B. Arrieta et al., Explainable Artificial Intelligence (XAI): Concepts, taxonomies, opportunities and challenges toward responsible AI (2019)

[Bar03] J. Barnes, *Porphyry: Introduction.* Oxford University Press UK (2003)

[Beg+20] T. Begley et al., *Explainability for fair machine learning* (2020)

[Bel59] W.A. Belson, Matching and Prediction on the Principle of Biological Classification. J. Roy. Stat. Soc. Ser. C **8**(2), 65–75 (1959)

[Bro+20] T.B. Brown et al., *Language models are few-shot learners* (2020). arXiv:2005.14165 [cs.CL]

[Bru+20] M. Brundage et al., *Toward trustworthy AI development: Mechanisms for supporting verifiable claims* (2020)

[CPC19] D.V. Carvalho, E.M. Pereira, J.S. Cardoso, Machine learning interpretability: A survey on methods and metrics. Electronics **8**(8), 832 (2019)

[Che+20] L. Chen et al., *Counterfactual samples synthesizing for robust visual question answering* (2020)

[CAL20a] Q. Chen, A. Allot, Z. Lu, Keep up with the latest coronavirus research. Nature **579**(7798), 193 (2020). ISSN: 1476-4687 (Electronic) 0028-0836 (Linking). https://doi.org/10.1038/d41586-020-00694-1. https://www.ncbi.nlm.nih.gov/pubmed/ 32157233.

[CAL20b] Q. Chen, A. Allot, Z. Lu, LitCovid: an open database of COVID-19 literature. Nucl. Acids Res. **49**(D1), D1534–D1540 (2020)

[Dar17] A. Dardagan, *Neoplatonic "Tree of Life" (Arbor Porphyriana: A diagram of logic and mystical theology)* (Mar. 2017). https://doi.org/10.31235/osf.io/g2qxe. osf.io/preprints/socarxiv/g2qxe

[DR20a] A. Das, P. Rad, Opportunities and challenges in explainable artificial intelligence (xai): A survey. Preprint (2020). arXiv:2006.11371

[DR20] A. Das, P. Rad, Opportunities and Challenges in Explainable Artificial Intelligence (XAI): A Survey. CoRR (2020). arXiv:abs/2006.11371. http://dblp.uni-trier.de/db/ journals/corr/corr2006.html#abs-2006-11371

[DK17] F. Doshi-Velez, B. Kim, *Towards a rigorous science of interpretable machine learning* (2017)

[Dos+19] F. Doshi-Velez et al., *Accountability of AI under the law: The role of explanation* (2019). arXiv:1711.01134 [cs.AI]

[DLH18] M. Du, N. Liu, X. Hu, Techniques for interpretable machine learning. CoRR (2018). arXiv:abs/1808.00033. http://dblp.uni-trier.de/db/journals/corr/corr1808.html# abs-1808-00033

[FP21] T. Feldman, A. Peake, *On the basis of sex: A review of gender bias in machine learning applications* (2021)

[Fer+20] X. Ferrer et al., *Bias and discrimination in AI: A cross-disciplinary perspective* (2020)

[Fis36] R.A. Fisher, The use of multiple measurements in taxonomic problems. Ann. Eugenics **7**(7), 179–188 (1936)

[GSC18] J. Garcia-Gathright, A. Springer, H. Cramer, *Assessing and addressing algorithmic bias - But before we get there* (2018)

[Gil+19] L.H. Gilpin et al., *Explaining explanations: An overview of interpretability of machine learning* (2019)

[He+15] K. He et al., *Delving deep into rectifiers: Surpassing HumanLevel performance on ImageNet classification* (2015). arXiv:1502.01852 [cs.CV]

[Hol+19] A. Holzinger et al., Causability and explainability of artificial intelligence in medicine. WIREs Data Mining Knowl. Discovery **9**(4), e1312 (2019). https://doi.org/10.1002/ widm.1312

[IG20] C. Intahchomphoo, O.E. Gundersen, Artificial intelligence and race: A systematic review. Legal Inf. Manag. **20**(2), 74–84 (2020)

[Isl+21] S.R. Islam et al., *Explainable artificial intelligence approaches: A survey* (2021)

[Isl+21b] S.R. Islam et al., Explainable artificial intelligence approaches: A survey. Preprint (2021). arXiv:2101.09429

[JMB20] D. Janzing, L. Minorics, P. Bloebaum, Feature relevance quantification in explainable AI: A causal problem, in *Proceedings of the Twenty Third International Conference on Artificial Intelligence and Statistics*, ed. by S. Chiappa, R. Calandra, vol. 108. Proceedings of Machine Learning Research (PMLR, 2020), pp. 2907–2916

[JSB20] M. Juric, A. Sandic, M. Brcic, *AI safety: state of the field through quantitative lens* (2020). arXiv:2002.05671 [cs.CY]

[KK62a] J.F. Kenney, E.S. Keeping, *Mathematics of statistics* (Princeton, 1962), pp. 252–285

[KH19] A. Khademi, V. Honavar, *Algorithmic bias in recidivism prediction: A causal perspective* (2019)

[Kle+18] J. Kleinberg et al., Human decisions and machine predictions. Q. J. Econ. **133**(1), 237– 293 (2018)

[LB20] H. Lakkaraju, O. Bastani, How do I fool you? Manipulating user trust via misleading black box explanations, in *Proceedings of the AAAI/ACM Conference on AI, Ethics, and Society* (2020), pp. 79–85

[Lak+19] H. Lakkaraju et al., Faithful and customizable explanations of black box models, in *Proceedings of the 2019 AAAI/ACM Conference on AI, Ethics, and Society* (2019), pp. 131–138

[Lan13] B. Lantz, *Machine learning with R*. Packt Publishing Ltd (2013)

[Lea18] S. Leavy, Gender bias in artificial intelligence: The need for diversity and gender theory in machine learning, in *GE '18* (Association for Computing Machinery, Gothenburg, Sweden, 2018), pp. 14–16. ISBN: 978-1-45-03573-88

[Lea+20] S. Leavy et al., *Mitigating gender bias in machine learning data sets* (2020)

[MGG01] D.H. Maister, C.H. Green, R.M. Galford, *The trusted advisor.* A Touchstone book (Free Press, 2001). ISBN: 978-0-74-32123-42

[Meh+19] N. Mehrabi et al., *A survey on bias and fairness in machine learning* (2019). arXiv:1908.09635 [cs.LG]

[MM72] R. Messenger, L. Mandell, A modal search technique for predictive nominal scale multivariate analysis. J. Am. Stat. Assoc. **67**(340), 768–772 (1972)

[MHS17] T. Miller, P. Howe, L. Sonenberg, *Explainable AI: Beware of inmates running the asylum Or: How i learnt to stop worrying and love the social and behavioural sciences* (2017)

[Mni+13] V. Mnih et al., *Playing Atari with deep reinforcement learning* (2013). arXiv:1312.5602 [cs.LG]

[MZR20] S. Mohseni, N. Zarei, E.D. Ragan, *A multidisciplinary survey and framework for design and evaluation of explainable AI systems* (2020)

[Mol19] C. Molnar, *Interpretable machine learning A guide for making black box models explainable* (2019)

[MCB20a] C. Molnar, G. Casalicchio, B. Bischl, *Interpretable machine learning – A brief history state-of-the-art and challenges* (2020)

[MCB20b] C. Molnar, G. Casalicchio, B. Bischl, *Interpretable machine learning—A brief history, state-of-the-art and challenges*. Preprint (2020). arXiv:2010.09337

[MS21] M. Moradi, M. Samwald, Post-hoc explanation of blackbox classifiers using confident itemsets. Expert Syst. Appl. **165**, 113941 (2021)

[MS63] J. Morgan, J. Sonquist, Problems in the analysis of survey data, and a proposal. J. Am. Stat. Assoc. **58**, 415–434 (1963)

[Mur+19] W.J. Murdoch et al., Interpretable machine learning: definitions, methods, and applications. Preprint (2019). arXiv:1901.04592

[Nil09] N.J. Nilsson, *The Quest for Artificial Intelligence*, 1st. (Cambridge University Press, USA, 2009). ISBN: 978-0-52-11229-37

[Nto+20] E. Ntoutsi et al., *Bias in data-driven AI systems – An introductory survey* (2020)

[Pea85] J. Pearl, A constraint - Propagation approach to probabilistic reasoning, in *Proceedings of the First Conference on Uncertainty in Artificial Intelligence, UAI'85* (AUAI Press, Los Angeles, CA, 1985), pp. 31–42. ISBN: 978-0-44-40058-7

[Pea94] J. Pearl, A probabilistic calculus of actions, in *Proceedings of the Tenth International Conference on Uncertainty in Artificial Intelligence* (Morgan Kaufmann Publishers, Seattle, WA, 1994), pp. 454–462. ISBN: 978-1-55-86033-28

[Phi+20] P. Phillips et al., *Four principles of explainable artificial intelligence (draft)* (2020)

[Pra+20] M. Prabhushankar et al., *Contrastive explanations in neural networks* (2020)

[Rai19] A. Rai, Explainable AI: From black box to glass box. J. Acad. Market. Sci. **48**, 137–141 (2019). https://doi.org/10.1007/s11747-019-00710-5

[RWC20] C. Rudin, C. Wang, B. Coker, The age of secrecy and unfairness in recidivism prediction. Harvard Data Sci. Rev. **2**(1), (2020). https://hdsr.mitpress.mit.edu/pub/7z10o269

[RHW86] D.E. Rumelhart, G.E. Hinton, R.J. Williams, Learning internal representations by error propagation, in ed. by D.E. Rumelhart, J.L. Mcclelland (MIT Press, 1986), pp. 318–362

[SSS20] S.A. Seshia, D. Sadigh, S. Shankar Sastry, *Towards verified artificial intelligence* (2020)

[Shl14] J. Shlens, *A tutorial on principal component analysis* (2014). arXiv:1404.1100 [cs.LG]

[Smi+88] J.W. Smith et al., Using the ADAP learning algorithm to forecast the onset of diabetes mellitus, in *Proceedings of the Annual Symposium on Computer Application in Medical Care* (American Medical Informatics Association, 1988), p. 261

[Tan+09] P. Tans et al., Trends in atmospheric carbon dioxide-Mauna Loa. Retrieved December **12**(2009), 2009 (2009)

[Tho19] T. Davidson, D. Bhattacharya, I. Weber, Racial bias in hate speech and abusive language detection datasets, in *Proceedings of the Third Workshop on Abusive Language Online* (Association for Computational Linguistics, Florence, Italy, 2019), pp. 25–35

[Tur95] A.M. Turing, *Computers & amp; thought* (MIT Press, 1995), pp. 11–35. Chap. Computing Machinery and Intelligence

[VDH20] S. Verma, J. Dickerson, K. Hines, *Counterfactual explanations for machine learning: A review* (2020)

[VM20] L. Vigano, D. Magazzeni, Explainable security, in *2020 IEEE European Symposium on Security and Privacy Workshops (EuroS&PW)* (IEEE, 2020), pp. 293–300

[WMR18] S. Wachter, B. Mittelstadt, C. Russell, *Counterfactual explanations without opening the black box: Automated decisions and the GDPR* (2018)

[Wan+20] A. Wang et al., *SuperGLUE: A stickier benchmark for general-purpose language understanding systems* (2020). arXiv:1905.00537 [cs.CL]

[WM94] I. Watson, F. Marir, Case-based reasoning: A review. Knowl. Eng. Rev. **9**(4), 327–354 (1994)

[WR20] C. Wolf, K. Ringland, Designing accessible, explainable AI (XAI) experiences. ACM SIGACCESS Accessibil. Comput., 1–1 (2020). https://doi.org/10.1145/3386296.3386302

[XRV17] H. Xiao, K. Rasul, R. Vollgraf, *Fashion-MNIST: A novel image dataset for benchmarking machine learning algorithms* (2017). arXiv:cs.LG/1708.07747 [cs.LG]

[Yua+21] H. Yuan et al., *Explainability in graph neural networks: A taxonomic survey* (2021)

[Zha19] J. Zhang, *Basic neural units of the brain: Neurons, synapses and action potential* (2019). arXiv:1906.01703 [q-bio.NC]

Chapter 2
Pre-model Interpretability and Explainability

This chapter discusses various ways of using pre-modeling explainability: a set of techniques aimed at gaining insights into a dataset to help build more effective models. Since any machine learning model is built from the data, understanding the content on which the model is based is imperative for explainability and interpretability. Many of these techniques that summarize, visualize, and explore data have existed for a long time. There have been some recent additions to improve the methods, especially with respect to scaling and performance. We will present exploratory data analysis with some well-known univariate and multivariate techniques to visualize the data. Time series data needs different transformations, visualizations, and analysis than structured data. Some of the common time series visualizations and data exploration techniques will be discussed next. Similarly, exploratory data analysis on unstructured data such as text needs special handling as compared to structured data. This will entail a discussion on some of the well-known EDA techniques common to many NLP tasks. We will then discuss some of the feature engineering techniques that help us get more insights for explainability. For our discussions, we will be using the diabetes dataset for structured data: LitCovid (for NLP data), Mauna Loa CO2 dataset (for time series data), and Fashion MNIST (for computer vision).

2.1 Data Science Process and EDA

Figure 2.1 shows a detailed view in the exploratory data analysis part of the data science process where univariate and multivariate analysis, visualizations, and feature engineering give insights into the data [Tuf11]. In real-world model development, this process of data cleansing, visualization, feature engineering, and modeling happens multiple times until the final model(s) meets the business criteria. From an explainability perspective, whether for diagnostics or interpretability,

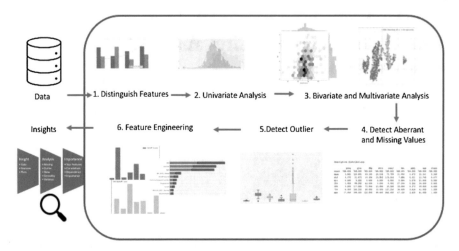

Fig. 2.1 Exploratory data analysis showing various techniques employed to convert data into insights

understanding the pre-model building process in a white-box manner is essential. This chapter will discuss exploratory data analysis and feature engineering in detail from an explainability standpoint.

2.2 Exploratory Data Analysis

Exploratory Data Analysis (EDA) can be defined as a collection of statistical techniques and visualization representations to get more insights from the data [Tuf11, Gho+18]. EDA was pioneered by J. Tukey in 1970s with the goal of diagnosis, hypothesis-free analysis, and laying the foundation stone to understand the whole story around the data [Tuk77]. EDA helps us understand the data distribution, the data quality, and the relationships that may exist between various features in the data. EDA helps in understanding the irregularities in the data and bias in the model, to some extent. Since data exist in various forms—images, texts, tabular, voice, graph, etc.—there are specific techniques for the specialized data formats.

2.2.1 EDA Challenges for Explainability

Some of the challenges posed by real-world data for interpretability while performing EDA are listed below.

1. Dimensionality: Many real-world datasets have very high-dimensionality or a large number of attributes or features [Liu et al., 2017a)]. One basic challenge this high-dimensionality poses is the increase in the computation cost to analyze and visualize proportionally. As the dimensionality grows, the visualization techniques that rely on capturing variations or projecting the data into lower dimensions become arduous due to the loss of the information [JT09].
2. Mixed-Types or Heterogeneity: In many industrial datasets, there will be features of different types such as categorical, continuous, ordinal, etc. Performing statistical analysis, especially on categorical features or doing exploration through sorting, results in performance challenges [JT09, Tuf11].
3. Missing Values and Outliers: Most real-world datasets typically show incompleteness with many missing or irregular values. Another common problem is the data having an outlier or atypical values caused by errors. Handling the missing values and outliers is a topic by itself and greatly impacts the model outputs [CBK09].
4. Anonymity and Privacy: Many features or attributes may contain classified or sensitive information that may not be accessible to every person unless their role in an organization permits it. Addressing privacy-based concerns is fast evolving as a critical topic in machine learning [Cri20].
5. Volume: Most contemporary datasets are large in size, often containing billions of data points, posing problems for traditional techniques that are memory-based. Thus, exploration techniques must rely on adopting methods such as approximation, sampling, or various aggregations [Bik19].

2.2.2 EDA: Taxonomy

There are many ways to categorize EDA, but we will go with the simplest one employed in the literature [Sel]. One feature at a time—univariate analysis–and multiple features together—multivariate analysis—form the first differentiating method to explore the data. Using graphical versus non-graphical approaches forms the second distinguishing method. The table below highlights all the four categories with some well-known techniques for each that will be explored in the concerned sections (Table 2.1).

2.2.3 Role of EDA in Explainability

These are some of the roles EDA plays in explainability and interpretability [Sel]:

- Identify characteristics such as the spread of the data, ranges, and missing values using univariate statistics.

Table 2.1 Taxonomy of
Common EDA techniques

	Graphical	Non-graphical
Univariate	Boxplots, density plots, violin Plots, etc.	Summary statistics
Multivariate	Pair plots, heatmaps, biplots, joint plots, radial charts, projection plots (T-SNE, MDS, etc.), and parallel coordinates plots	Correlation analysis

- Overcome issues of outliers in modeling using outlier and distribution analysis [CBK09].
- Help gain insights and impute missing values [Gra09].
- Help project the data in lower dimensions for ease of visualization and provide insights into local and global data patterns using multivariate visualization techniques.
- Enable understanding of relationships between one or more features and their correlations which are useful for local and global pattern analysis.
- Validate hypothesis of a scenario by an expert.
- Validate a new feature constructed using one or more existing features by looking at its local and global interactions.

2.2.4   Non-graphical: Summary Statistics and Analysis

One of the necessary steps in any data science process, especially for explainability, is getting summary statistics from raw or transformed data. The summary statistics generally includes details for every raw feature such as data type, unique values, missing values, duplicate rows, and most frequent values, among others. Working with the domain expert and mapping these features onto categorical (nominal), continuous (numeric), temporal, etc. is a crucial step before modeling. Analyzing the data through statistical measures such as ranges, quartiles, standard deviations, kurtosis, and skewness can further our understanding of the distribution and bias.

The goal of univariate summary statistics is to explain each feature's sample distribution and thus extrapolate it for the population distribution. A domain or subject matter expert can validate this information for sampling bias and other types of errors.

2.2.4.1 Tools and Libraries

There are many Python libraries with different capabilities for performing EDA. Some of them are statsmodel, Holoviews, lens, pandas-summary, pandas-profiling, and SweetViz [SP10, Ste+19, tea20, Bru19, Ber19]. Most of these packages provide necessary statistical analysis detailed in Sect. 2.2.4. In this chapter we will use pandas and SweetViz as our EDA tool to analyze the Pima Indian diabetes dataset discussed in Chap. 1.

2.2.4.2 Summary Statistics and Analysis

The summary statistics from basic pandas *describe* is given in Fig. 2.2 with counts and distribution statistics. For a given dataset $\{x_i\}_{i=1}^{N}$, the following statistical measures for feature characteristics can be defined:
   the k-th central moment:

$$\mu_k \cong \frac{1}{N} \sum_{i=1}^{N} (x_i - \mu_i)^k \tag{2.1}$$

Mean:

$$\mu_1 \cong \frac{1}{N} \sum_{i=1}^{N} x_i \tag{2.2}$$

Variance:

$$\mu_2 \cong \frac{1}{N} \sum_{i=1}^{N} (x_i - \mu_i)^2 \tag{2.3}$$

Standard Deviation:

$$\mu_2 = \sqrt{\mu_2} \tag{2.4}$$

Skew:

$$\gamma = \frac{\mu_3}{\sigma^3} \tag{2.5}$$

Kurtosis:

$$\kappa = \frac{\mu_4}{\sigma^4} \tag{2.6}$$

Excess Kurtosis:

$$\kappa_e = \kappa - 3 \tag{2.7}$$

Tailing:

$$\tau = \frac{\mu_5}{\sigma^5} \tag{2.8}$$

Figure 2.3 shows the skew statistics for all the features.

> The skew statistics showing features with large positive and negative skew indicates the potential need for some normalization techniques such as standardization or min–max scaling based on the modeling algorithm.

The summary statistics, as shown in Fig. 2.4, give a detailed report for the diabetes dataset in terms of duplicates present in the data and the number of categorical, continuous, and other features such as text present. The report provides essential statistical measurements such as unique values, missing values, and duplicates for all the features. For every continuous feature, it further reports statistical properties that measure centrality and spread like mean, median, inter-quartile ranges, and variance, to name a few, as shown in Fig. 2.4. Categorical features are best statistically represented with frequency by category.

```
Descriptive distributions:

            preg      gluc       dbp      skin     insul       bmi      pedi       age     class
count    768.000   768.000   768.000   768.000   768.000   768.000   768.000   768.000   768.000
mean       3.845   120.895    69.105    20.536    79.799    31.993     0.472    33.241     0.349
std        3.370    31.973    19.356    15.952   115.244     7.884     0.331    11.760     0.477
min        0.000     0.000     0.000     0.000     0.000     0.000     0.078    21.000     0.000
25%        1.000    99.000    62.000     0.000     0.000    27.300     0.244    24.000     0.000
50%        3.000   117.000    72.000    23.000    30.500    32.000     0.372    29.000     0.000
75%        6.000   140.250    80.000    32.000   127.250    36.600     0.626    41.000     1.000
max       17.000   199.000   122.000    99.000   846.000    67.100     2.420    81.000     1.000
```

Fig. 2.2 Descriptive statistics using pandas

Fig. 2.3 Measuring skew of the distribution for all the features using pandas

```
Skew of features:

preg      0.902
gluc      0.174
dbp      -1.844
skin      0.109
insul     2.272
bmi      -0.429
pedi      1.920
age       1.130
class     0.635
```

Many summary statistics provide correlation analysis such as Pearson's correlation for a feature with all other features, including the target. This helps to explain inter-dependencies between the features at a very basic level. Figure 2.5 shows a detailed association analysis for the feature *SkinThickness*. Multivariate analysis is used to compute statistics that show the interaction between features or between a feature and the target [And94, MKB79]. Pairwise covariances and correlations across continuous features in the form of either cross-tabulation or matrix are a common non-graphical analysis. A similar correlation/covariance analysis is also visualized using heatmaps and will be discussed in the next section.

> Correlation and association analysis helps in explaining the interdependence among the features. This can help in either simplifying the feature space (removing the redundant features) or improving the interpretability of the machine learning model.

A pivot table is a descriptive summary statistics technique that enables the user to analyze statistics such as sums, means, standard deviations, etc. by groups for diagnostics or analysis. In machine learning scenarios, one employs the **pivot** with

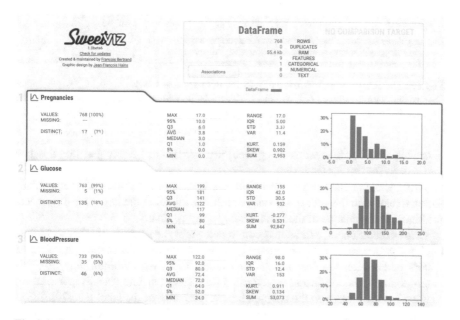

Fig. 2.4 Snapshot report generated by SweetViz EDA analysis on the diabetes dataset. It gives statistics about the entire dataset and for each feature such as *Pregnancies*, *Glucose*, and *BloodPressure* the distributions

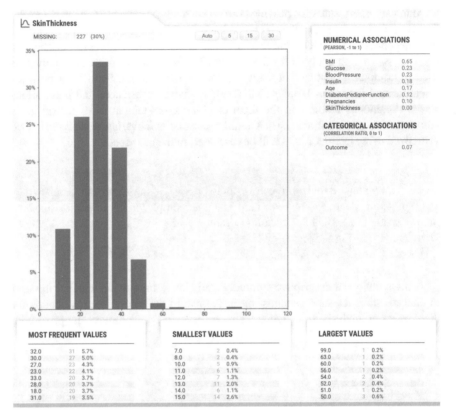

Fig. 2.5 The feature *SkinThickness* analysis for the statistical metrics, associations with other features and with the outcome. It shows a high correlation with *BMI* and a low correlation with *Pregnancies*

either the labels or interesting categorical features that can easily partition the data into groups for further analysis.

In Fig. 2.6, the class or the label *Outcome* is used as the pivot with mean values for each feature in the data as a measure of the distribution; the larger the difference, the higher is the chance of the feature being a discriminative one.

Cross-tabulation, also known as contingency table analysis, is another technique used for understanding the relationship between the features mapped as discrete factors or between the features and the label. The most common way of using cross-tabulation is converting hypothesis into discrete factors (mostly per feature) to visualize the impact on the subpopulations per factor or the label [Hab72].

Figure 2.7 shows one such analysis on how domain-specific factors, e.g., in the medical-based instance, indicators such as being obese (*BMI* > 30), high insulin (*Insulin* > 166), high glucose (*Glucose* ≥ 140), high skin thickness (*SkinThickness* ≥ 27), and considered young (*Age* < 30), can be used to partition the data

Mean values of features by class:

class	age	bmi	dbp	gluc	insul	pedi	preg	skin
0	31.190	30.304	68.184	109.980	68.792	0.43	3.298	19.664
1	37.067	35.143	70.825	141.257	100.336	0.55	4.866	22.164

Fig. 2.6 Pivot table showing mean for each feature by class

age_young	0								1							
bmi_obese	0				1				0				1			
skinthickness_high	0		1		0		1		0		1		0		1	
glucose_high	0	1	0	1	0	1	0	1	0	1	0	1	0	1	0	1
Outcome																
0	0.587302	0.186047	0.518987	0.22807	0.74026	0.409091	0.782609	0.625	0.754098	0.333333	0.809091	0.3	0.956522	0.727273	0.9	0.75
1	0.412698	0.813953	0.481013	0.77193	0.25974	0.590909	0.217391	0.375	0.245902	0.666667	0.190909	0.7	0.043478	0.272727	0.1	0.25

Fig. 2.7 Cross tabulation

and understand the risk impact [Gri18]. Cross-tabulation also helps in validating hypothesis and explainability from the data perspective. Often, such factors, e.g., obesity or high glucose, can be added as new features for capturing explicit impact from the subject matter expert perspective while modeling.

Observations:

- Figure 2.2 shows that features *BloodPressure* for measuring diastolic BP, *SkinThickness* for triceps skin fold thickness, *Insulin* for 2 h serum insulin, and *BMI* for *BMI* have many 0 values, which is not biologically possible, and hence represent missing or aberrant values.
- Further analysis shows that the 25th percentile of the features *Insulin* and *SkinThickness* is 0. This means that at least a quarter of the data points are missing these features, so we need some form of missing value imputation technique to be employed for modeling.
- The skew statistics in Fig. 2.3 shows negative skew for features *BloodPressure* and *BMI*, while it shows positive skew for features *Insulin*, *Age*, and *DiabetesPedigreeFunction*.

- The pivot table as in Fig. 2.6 of the features against label shows meaningful differences in the mean values indicating that these features can be useful in discriminating the class; however, missing values coded as zero can artificially create these differences in the means and need to be verified again after the imputation process.
- The cross-tabulation results validate many of the medical hypotheses that in younger women (age < 30) the risk of diabetes almost doubles from 41 to 81% with high glucose (glucose ≥ 140) and stays high with other risk factors such as high skin thickness (SkinThickness ≥27). Similarly, in older women (age > 30), the risk with high skin thickness and high glucose almost triples the chances of diabetes from 24 to 70%.

2.2.5   Graphical: Univariate and Multivariate Analysis

Graphical techniques complement the non-graphical methods in giving visual and qualitative explanations to the sample distribution, local or global interactions, and inter-relationships between the features.

2.2.5.1   Tools and Libraries

Some well-known open-source Python libraries that are useful for univariate and multivariate analysis are Matplotlib, Plotly, Seaborn, and Bokeh [Hun07, Wt20, Bok20, Inc15].

2.2.5.2   Univariate Analysis

Univariate visualization is one of the essential steps in the EDA process. Univariate visualization mostly helps explain and understand the dataset when there are low to medium features (less than 20). This section highlights some well-known univariate visualization methods used in EDA to explain, debug, and validate the data or the models.

Boxplots

Boxplots and many of its variants are excellent tools to visualize the central tendency (median), symmetry (location of median and whisker length), and the outliers (outside the IQR range) as shown in Fig. 2.8

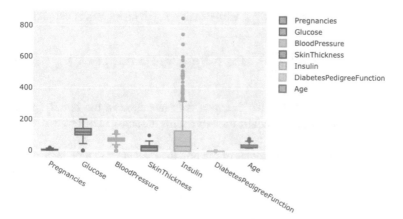

Fig. 2.8 Boxplot capturing distributions for all the features in the diabetes dataset

Boxplots use robust statistics (median and IQR) and give useful and reliable statistics based on the sample. The location of median/mean in the box is a measure of skewness. The length of the boxplot measures the spread, while the length of the whiskers measures the tail length of the distribution. The way to determine which data points are the real outliers and how to treat them is a complex matter and has real implications on the model [Agg16]. A disadvantage of boxplots is that they do not show multimodality in the distribution.

Distribution Plots

Distribution plots show combined information in the histograms: kernel density estimation (KDE) and the rug plot. Histograms show the distribution of the sample based on a discrete number of bins and frequency of data in them, while KDE does the smoothing based on the Gaussian kernel. The accuracy of histograms depends on the parameter choice of the number of bins while that of KDE depends on the bandwidth parameter. Histograms and KDE give a good visual explanation of the distribution, especially the central tendency, skew, and multimodality (Fig. 2.9). One has to observe caution, especially when there are discrete features or natural bounds associated with features (e.g., the number of pregnancies in the diabetes dataset).

For histograms, if x_0 is the origin, and the length of a bin is h, then a bin $B_j(x_0, h)$ is given by

$$B_j(x_0, h) = [x_0 + (j-1)h, x_0 + jh), j \in \mathbb{Z}] \qquad (2.9)$$

If for a given i.i.d. dataset $\{x_i\}_{i=1}^{n}$ with density f, the histogram is defined as

$$\hat{f}_h(x) = n^{-1}h^{-1} \sum_{j\in\mathbb{Z}} \sum_{i=1}^{n} \mathbf{I}\{x_i \in B_j(x_0, h)\}\mathbf{I}\{x \in B_j(x_0, h)\} \tag{2.10}$$

where $\mathbf{I}\{x_i \in B_j(x_0, h)\}$ is the count of data instances in the bin $B_j(x_0, h)$ and $\mathbf{I}\{x \in B_j(x_0, h)$ represents localization of counts around x. The optimal value of h is generally $h = (24\sqrt{\pi}/n)^{1/3}$.

By supposing x to be the center of the bin, the histogram can be rewritten as

$$\hat{f}_h(x) = n^{-1}h^{-1} \sum_{i=1}^{n} \mathbf{I}\left(|x - x_i| \leq \frac{h}{2}\right) \tag{2.11}$$

If we set $K(u) = \mathbf{I}(|u| \leq \frac{1}{2})$, then the above equation can be rewritten as

$$\hat{f}_h(x) = n^{-1}h^{-1} \sum_{i=1}^{n} K\left(\frac{x - x_i}{h}\right) \tag{2.12}$$

Equation 2.12 is the general form of kernel estimator. Different kernels create different shapes of the estimated density. A Gaussian kernel is given by

$$K(u) = \frac{1}{\sqrt{2\pi}} \exp\left(\frac{-u^2}{2}\right) \tag{2.13}$$

Distribution plots can guide many probabilistic classifiers over estimators' choice for the probability distribution of the feature. Distribution plots also help in validating the assumptions made of modality in the classifiers. Modes are the peaks in the histograms. Distribution plots can also be used for imputing the missing values for the features.

Violin Plots

Violin plots combine many aspects of boxplots and distribution plots and have the joint advantage of both of them in one plot. Figure 2.10 shows violin plots for all the features in the diabetes dataset.

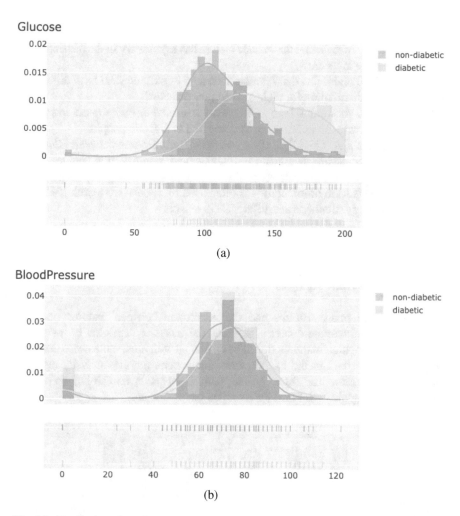

Fig. 2.9 Distribution of two features *Glucose* and *BloodPressure* in the dataset using distribution plot. (**a**) Feature *Glucose* distribution. (**b**) Feature *BloodPressure* distribution

---

Observations:

- Figure 2.8 shows many features such as *Glucose*, *BloodPressure*, *Insulin*, and *SkinThickness* containing quite a few outliers.
- The *Insulin* feature shows clear asymmetry in the distribution with a large left skew.

- Figure 2.10 reveals that the *Insulin* feature has asymmetry in the distribution with a large left skew.
- Figure 2.10 highlights that features such as *Glucose* and *BMI* have different mean and distribution for diabetes and non-diabetes.
- Figure 2.10 shows that both *SkinThickness* and *BloodPressure* are lower in the people who are not diabetic as compared to diabetic people.
- Figure 2.9a and b highlight the distributional differences between *Glucose* and *BloodPressure* features for diabetic and non-diabetic people. The estimation of missing values conditioned on the outcome would be a better choice. This form of imputation is often used for supervised learning where the feature distributions are different for each class.

2.2.5.3   Multivariate Analysis

When there are many features and the interaction between features needs to be analyzed, multivariate analysis provides the necessary insights by using the techniques highlighted in the section. The challenge with most multivariate analysis is projecting the data in the two or three dimensions for the analysis, resulting in some information loss. Multivariate analysis helps in explaining both local interactions and global relationships based on the technique.

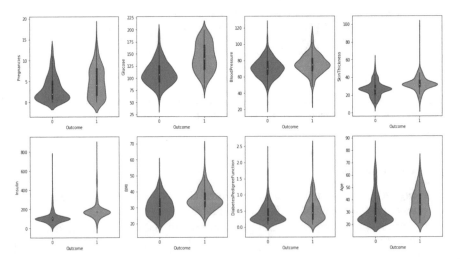

Fig. 2.10 Violin plots capturing distributions for all the features

Joint Distribution Plots

The bivariate plots, as shown in Fig. 2.11a and b, are useful to see the interactions and coupling between two continuous features. Not only does it help in confirming the assumptions for modeling, but also in identifying the outliers and providing explanations for the interactions between the features. Joint distribution plots are good for explaining local interactions and behaviors but not global properties.

The joint behavior of two features X and Y is fully captured in the joint probability distribution. For a continuous distribution,

$$E[X^m Y^m] = \int \int_{-\infty}^{\infty} x^m y^m f_{XY}(x, y)dxdy \tag{2.14}$$

For discrete distribution,

$$E[X^m Y^m] = \sum_{x \in S_x} \sum_{y \in S_y} x^m y^m P(x, y) \tag{2.15}$$

Observations:

- Figure 2.11a and b plots bivariate distribution for two features for only one class, i.e., only diabetic samples. This helps to understand the sampling bias as well as the explanation for relationship between these features.
- Most diabetic people between the *Ages* 25 and 60 had *BMI* between 25 and 45. That is, the relationship of an overweight condition or obesity to age group can be explained through this joint plot.
- Most diabetic people between the *Ages* 20 and 60 had *BloodPressure* between 50 and 90.

Heatmaps

A heatmap is a visualization of data in which the association between the features is colored based on correlations. One should avoid highly correlated variables when creating models because they capture the same occurrence and create "noise" or inaccuracy. Heatmap visualization is a great tool to aid the analyst when there is a large volume dataset that is not high-dimensional (less than 10). As shown in Fig. 2.12, both positive and negative correlations between every pair of features (including labels) give a quick summary of strong associations. Correlation

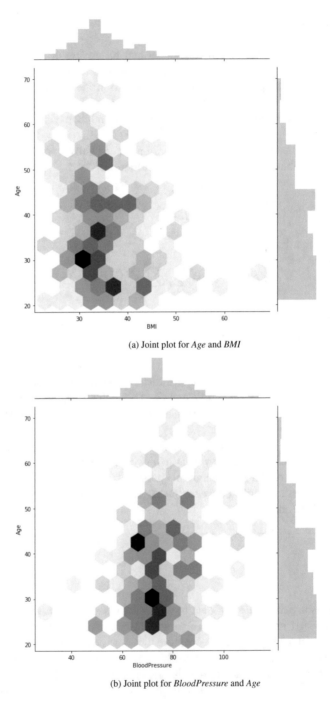

(a) Joint plot for *Age* and *BMI*

(b) Joint plot for *BloodPressure* and *Age*

Fig. 2.11 Joint distribution plots for diabetic patients only. (**a**) Joint plot for *Age* and *BMI*. (**b**) Joint plot for *BloodPressure* and *Age*

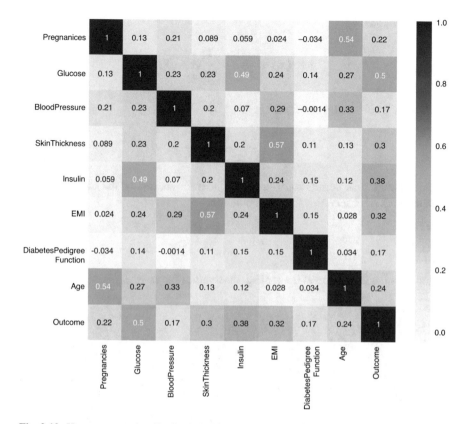

Fig. 2.12 Heatmap capturing distributions for all the features

analysis assumes that all the observations are independent of one another and the linear relationships between them may affect the analysis. Heatmaps are good for explaining local interactions and behaviors but not global properties.

Covariance between two features X and Y is given by

$$Cov(XY) = \frac{1}{n-1}\sum_{i=1}^{n}(x_i - \bar{x})(y_i - \bar{y})$$

(2.16)

Variance is given by

$$Var(X) = \frac{1}{n-1} \sum_{i=1}^{n} (x_i - \bar{x})^2 \qquad (2.17)$$

Pearson's correlation coefficient is given by

$$\rho_p = \frac{Cov(X, Y)}{\sqrt{Var(X)Var(Y)}} \qquad (2.18)$$

Spearman's rank correlation is a nonparametric measure of the correlation that uses the rank of observations in its calculation.

$$\rho_s = 1 - \frac{\sum d_i^2}{n(n^2 - 1)} \qquad (2.19)$$

where d_i is the difference in the ranked observation for $x_i - y_i$.

> Covariance and correlation are measures of linear dependence. Zero covariance or correlation does not imply independence. Negative or positive covariance or correlation is implied through slopes of scatter plots. Fisher's Z-transform helps us in testing hypotheses on correlation.
>
> It is important to note that various conditions, such as outliers, unequal variances, nonnormality, heteroscedasticity, and nonlinearities can disproportionately impact the Pearson correlation coefficient and the analysis. As a result of these problems, the Spearman correlation coefficient, based on the data ranks rather than the actual data, is possibly a better choice for examining the relationships between variables [Scho+2018].

Observations:

- *Glucose, Age, BMI,* and *Insulin* are the most correlated with the label *Outcome*.
- *BloodPressure* and *DiabetesPedigreeFunction* have a small correlation with the label *Outcome*.
- *BloodPressure* and *Insulin* have very little correlation with the label *Outcome*.
- The following features show a good correlation: *Age* and *Pregnancies, BMI* and *SkinThickness,* and *Glucose* and *Insulin*.

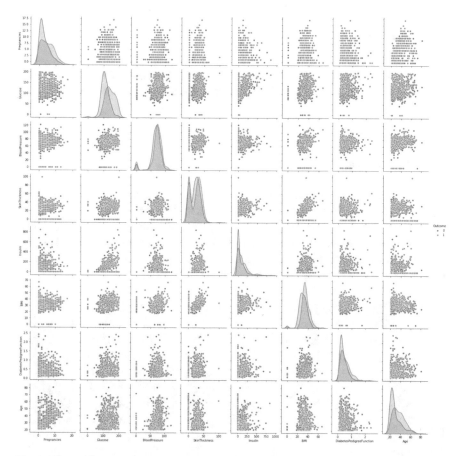

Fig. 2.13 Pair plots capturing distributions for all the features

Pair Plots

Pair plots, similar to heatmaps, capture the interaction between two continuous features but show scatter plots and density estimations between the features as shown in Fig. 2.13. Pair plots are good for explaining local interactions and behaviors but not global properties.

Observations:

- As seen in Fig. 2.13, feature pairs *BMI and SkinThickness*, *Pregnancies and Age*, and *Glucose and Insulin* seem to have positive linear relationships.

Parallel Coordinate Plots

Parallel coordinate visualization visualizes high-dimensional data in two-dimensional space. Vertical axes parallel to each other represent the features and the data element as a single line traversing through all the vertical axes. This two-dimensional visualization of a high-dimensional data is the primary advantage of the parallel coordinate plots. It becomes easy to understand the trend shown, especially when the label for each data is different [Eds03].

The order of the axes in the parallel coordinates plot can impact the way one analyzes the data. The main reason for this is that the relationships between adjacent features are easier to observe than non-adjacent features. One disadvantage of the parallel coordinates plots is that they can become over-cluttered and indecipherable in a large dataset. There are techniques such as "brushing" that isolate sections by filtering out the noise [Eds03]. Outliers are visible as outlying polygon curves. Various parameters such as normalization techniques to get all the features in the same range also impact the visualizations.

Observations:

- Figure 2.14 shows parallel coordinate plots for all the features and datasets with two different normalization techniques. Figure 2.14a uses standardization method to scale the features, while Fig. 2.14b uses $l2$ technique of weighting the features between 0 and 1. We see that the method of normalization has an impact on the plots and analysis.
- Figure 2.14a with the standardization form of preprocessing shows the overly cluttered plot and not much can be made out of it.

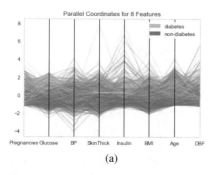
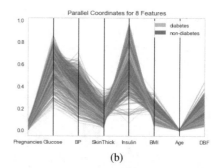

(a)   (b)

Fig. 2.14 Parallel coordinates plots with different normalization methods. (**a**) Parallel coordinates plot using standardization preprocessing. (**b**) Parallel coordinates plot using *l2* as preprocessing

- Figure 2.14b with *l2* normalization is clearer and shows some patterns of how instances having *Glucose*, *BloodPressure*, and *Insulin* are clustered together for diabetes and non-diabetes evident from the blue and green braids.

Projection Plots

Many well-known techniques project high-dimensional data in 2D or 3D plots useful for understanding local and global interactions. This section will discuss a few of the useful ones, especially from the explainability standpoint. Principal Component Analysis (PCA) is a statistical technique used to reduce the features by creating new features from important features that capture the most information about the dataset [SGM10]. The new features, known as the principal components, are the linear combinations of the existing features. The linear combination that captures the highest variance is the first principal component, the combination that captures the second highest variance is considered the second principal component, and so on.

If $\mathbf{X}$ is an $n \times p$ data matrix with n instances and p features, such that sample mean for the features is zero, the PCA transformation reduces the size to l from p through weights $\mathbf{w} = (w_1, \ldots, w_p)_{(k)}$, such that each row vector $\mathbf{x}_i$ maps to a new principal component matrix $\mathbf{T}_l = \mathbf{X}\mathbf{W}_l$ and element-wise is given by

$$t_{k(i)} = \mathbf{x}_{(i)} \cdot \mathbf{w}_{(k)} \; for \; i = 1, \ldots, n \; and \; k = 1, \ldots, l \qquad (2.20)$$

The principal component matrix $\mathbf{T}$ has columns that form an orthogonal basis for the l features that have been decorrelated. This can be achieved by an objective that preserves the maximum variances in the original data while minimizing the error $||\mathbf{TW}^T - \mathbf{T}_l\mathbf{W}_l^T||_2^2$.

PCA removes the correlations among the features and becomes an efficient tool for modeling techniques that rely on this feature. PCA also helps project the high dimension data in the lower dimension and captures many global and local interactions. There is a loss of interpretability in the data visualization concerning the original features as PCA has transformed the data into a new space. PCA is not scale invariant and the choice of scaling technique will have an impact on visualization and analysis.

A biplot is a combination of PCA and loading plot. Loading plot shows how the feature vectors influence the principal components in their magnitude and direction [Gab71].

The visualization of loading plots, especially the feature vectors' angles, gives an insight into the correlation characteristics. If two vectors are close, forming a small angle, they represent the two features positively correlated. If they are at 90, there is little to no correlation between them. When they diverge and form a large angle (close to 180), they are most negatively correlated.

The t-SNE algorithm has become a widespread technique in machine learning as it creates compelling two-dimensional projections for the data with a large number of features (hundreds or even thousands) [MH08]. The t-SNE algorithm is a non-linear transformation from an input space to a feature space.

The t-SNE algorithm maps the distance between two instances to the conditional probabilities representing similarities. Thus the conditional probability $p_{j|i}$ represents the probability of a data instance x_i picking another instance x_j as its neighbor with density measured by a Gaussian centered at x_i and is given by

$$p_{(j|i)} = \frac{\exp(-||x_i - x_j||^2/2\sigma_i^2)}{\sum_{k\neq i} \exp(-||x_k - x_j||^2/2\sigma_i^2)} \tag{2.21}$$

In the transformed space, the new conditional probability $q_{j|i}$ is given by

$$q_{(j|i)} = \frac{\exp(-||y_i - y_j||^2)}{\sum_{k \neq i} \exp(-||y_k - y_j||^2)} \tag{2.22}$$

Using stochastic gradient descent over y elements, the divergence between the two distributions can be minimized using the cost function C given by

$$C = \sum_{i,j} p_{(j|i)} \frac{p_{(j|i)}}{q_{(j|i)}} \tag{2.23}$$

The t-SNE algorithm has a tunable parameter known as "perplexity," enabling the user to balance the data's local and global aspects. The perplexity parameter is an approximation of the number of close neighbors each data element has. The learning rate and the number of iterations to train on the data impact the visualization. The t-SNE algorithm's concept of distance gets mapped to the regional density variations in the dataset. Thus, it generally expands the dense clusters in the data and contracts the sparse ones. Another fundamental aspect of the t-SNE plot is that the distances between well-separated clusters may not mean anything. The presence of clusters based on the perplexity parameter is what one should aim to see. The practical aspect is to plot t-SNE for various perplexities and observe the shapes and clusters for local and global interactions.

Isometric mapping (Isomap) is a non-linear dimensionality reduction method based on the spectral theory. Isomap maps the data from higher dimension to lower dimensions by preserving geodesic distances [TSL00]. The geodesic distances between two points use the graph distance between them, and hence it correctly approximates the close points as neighbors. Geodesic distances are better than the Euclidean distances in non-linear manifolds. Isomap uses the geodesic distance to create a similarity matrix for eigenvalue decomposition.

Isomap uses local information (geodesic distances from neighborhoods) to create a global similarity matrix. Isomap captures both the global and the local structure of the dataset in the low-dimensional projections.

Observations:

- Figure 2.15 with PCA does not show a good separation of diabetes and non-diabetes data. *This may indicate that linear mapping in the modeling may not be able to separate the classes.* However, the vectors for features *BMI*, *SkinThickness*, and *Insulin* are closer indicating correlations between them. Also, the features *Age* and *Pregnancies* indicate correlation by virtue of having angle less than 90.
- Figure 2.16 shows t-SNE plots for various values of perplexities. At *perplexity* 5 or 30, there is a good separation of the data into 4 clusters of diabetes and non-diabetes. *This shows that a non-linear mapping of features can yield a better classification.*
- Figure 2.17 shows Isomap projections of the features in 2D with good separation between diabetes and non-diabetes samples.

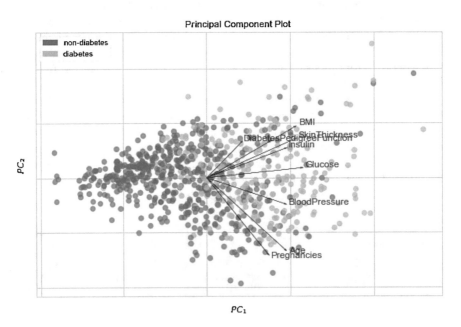

Fig. 2.15 Biplot capturing distributions for all the features

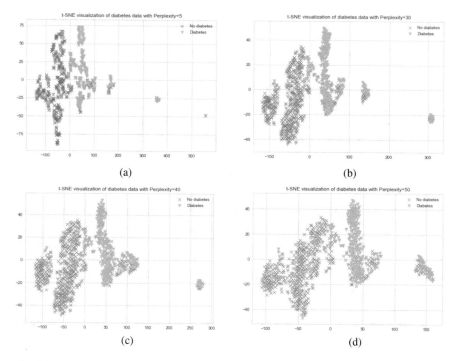

Fig. 2.16 t-SNE showing different plots for different values of perplexities. (**a**) t-SNE with perplexity = 5. (**b**) t-SNE with perplexity = 30. (**c**) t-SNE with perplexity = 40. (**d**) t-SNE with perplexity = 50

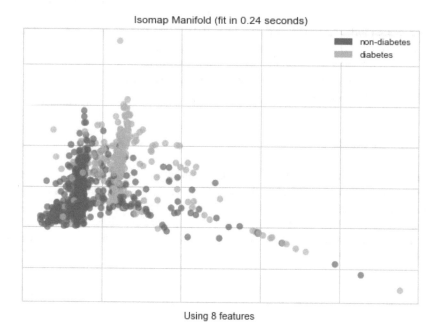

Fig. 2.17 Isomap projection of two features: *Glucose* and *BloodPressure* in the dataset

2.2.6   EDA and Time Series

In this section, we will describe techniques involving univariate time series data. Many traditional statistical forecasting methods such as Error Trend Seasonal (ETS), ARIMA, and SARIMA have high interpretability and explain the time series behavior based on the parameters and results [HM82, BD96, HA18b]. The time series EDA helps understand the data and guides in the choice of these interpretable models. Below are some of the advantages of performing EDA on time series data especially around interpretability.

1. It helps in identifying the missing values so that they can be either dropped or replaced by imputation. Imputing the missing values for time series is generally done using the forward fill or the backward fill or other interpolation techniques.
2. Analyzing the trend, seasonality, and cyclic behavior in the time series will help modeling strategy, especially the Error Trend Seasonal (ETS) models.
3. Many time series modeling algorithms like ARIMA are not robust to outliers. Seasonality may sometimes introduce outliers in time series, and hence careful treatment of outliers becomes imperative. Outlier analysis also influences model performance metrics. Root Mean Squared Error (RMSE) is not robust to outliers as compared to Mean Absolute Percentage Error (MAPE) [Jon80, BD91].
4. Statistical stationarity in the time series should have a constant mean, variance, and little autocorrelation at all lags (correlation with past values). The stationarity analysis guides the model choice as specific models, e.g., non-stationary time series, cannot be modeled using ARIMA but with ETS.
5. Structural breaks or shifts are abrupt changes in the trend that persists for a more extended period than outliers. The presence of such breaks and understanding them will be useful in modeling choices.

2.2.6.1   Resampling

Resampling the time series data refers to increasing or decreasing the frequency of the observations to explore the underlying behavior. Upsampling to a higher frequency involves interpolating new observations from existing ones, while down-sampling involves aggregating higher frequency observations. Resampling becomes inevitable if the prediction time frequencies are more or less frequent than the observations or the training data.

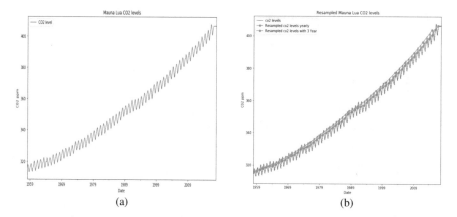

Fig. 2.18 Resampling plots for Mauna Loa CO_2 levels. (**a**) CO_2 level sampled at monthly rate. (**b**) Resampling CO_2 at different frequencies

---

Observations:

- Figure 2.18b indicates the resampling done using a year start as the frequency, as compared to monthly, completely removes the periodic behavior and captures the upward trend.
- There is a similarity between 3-year resampling and yearly sampling, indicating that 1-year observation frequency captures the trend.

2.2.6.2 Seasonality and Trend Analysis

The decomposition process is fundamental for studying time series data and exploring historical changes over time. Decomposition helps visualize the systematic and non-systematic components, i.e., the components to model and the noise part that cannot be modeled. Time series decomposition involves splitting a series into a combination of level, trend, seasonality, and noise components.

- The level is the mean or the average value in the time series.
- The trend is the rate of increasing (or decreasing) value in the time series (T_t).
- The seasonality is the short-term cyclic changes that happen (S_t).
- The noise or the residuals are randomness that is not possible to model (R_t).

Additive decomposition of the time series data y_t can be written as

$$y_t = T_t + S_t + R_t \tag{2.24}$$

Alternatively, a multiplicative decomposition would be written as

$$y_t = T_t \times S_t \times R_t \tag{2.25}$$

The additive decomposition is relevant when the seasonal variations or variations around the trends do not vary with the time series level. When seasonal variations or variations around the trends are proportional to the time series level, then multiplicative decomposition is more appropriate.

Observations:

- Figure 2.19 highlights that the central part of the series is the upward trend of the moving average and periodic seasonal pattern every year. The noise or the residual part is minimal, and we can neglect it in the modeling.

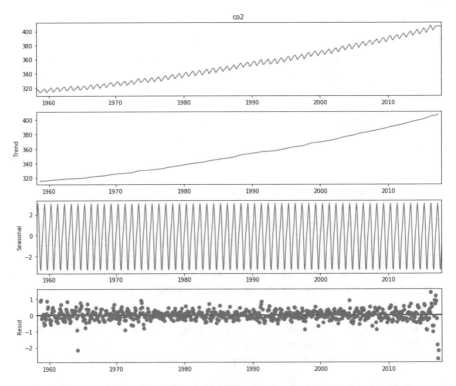

Fig. 2.19 Decomposition using additive method showing the trend, seasonal, and residual parts

2.2.6.3 Autocorrelation, Stationarity, and Differencing

Autocorrelation measures linear relationships between the lagged values in the time series (Fig. 2.20). High autocorrelation means that the observation point is more dependent or explained by another point (default is to measure consecutive observations or with a lag of 1).

A time series that has a trend (increasing or decreasing) or seasonality is not stationary. The presence of short-lived cycles may make it look non-stationary, but if the cycles have predictable patterns, it can be stationary.

> Another reason for EDA to understand stationarity is that many sample statistics such as means, variances, and covariances are useful as descriptors of future behavior only if the series is stationary.

If the time series shows a trend, it may be possible to stationarize it by de-trending, i.e., by fitting a trend line and subtracting it out before fitting a model. Such a time series is said to be trend-stationary. If the mean, variance, and autocorrelations of the time series are changing in time even after de-trending, it may imply that the variations between periods or between seasons may remain constant. Such a time series is said to be difference-stationary.

Transformations such as differencing (first difference) that computes the difference between two consecutive observations in the time series can make non-stationary time series stationary and hence enable interpretable models like ARIMA.

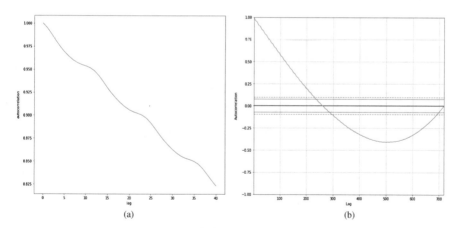

Fig. 2.20 Autocorrelations using both statsmodels (small value lags) and pandas (large value lags). (**a**) Autocorrelation using statsmodels. (**b**) Autocorrelation using pandas

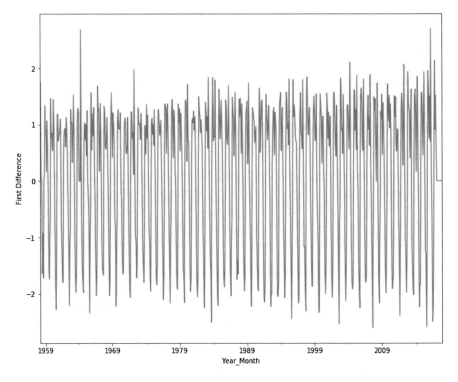

Fig. 2.21 First difference plot for Mauna Loa data

Observations:

- Figure 2.21 shows that de-trending or first differencing removes the trend completely and there is periodicity.
- Figure 2.22a and b shows that autocorrelation with differencing has clear yearly periodicity, i.e., every 12 months.

2.2.7   EDA and NLP

In this section, we will explore typical exploratory data analysis performed on text corpus during various NLP tasks such as classification, categorization, and summarization, among others. Exploratory data analysis involves text statistics visualization such as word or phrase frequencies, sentence length distribution

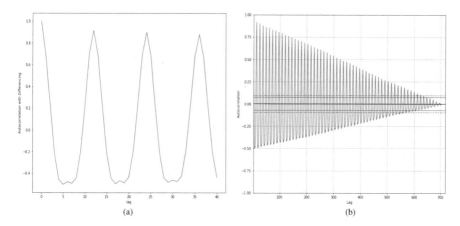

Fig. 2.22 Autocorrelations using both statsmodels (small value lags) and pandas (large value lags). (**a**) Autocorrelation after first differencing using statsmodels. (**b**) Autocorrelation after first differencing using pandas

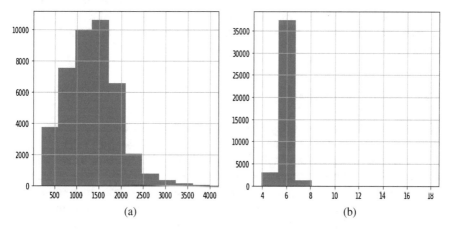

Fig. 2.23 Text Corpus statistics for LitCovid dataset. (**a**) Word distribution. (**b**) Average word length distribution

analysis, and topic analysis with relevant words defining it through unsupervised techniques [KLW19a].

2.2.7.1 Text Corpus Statistics

Many machine learning models are biased by the length of the sentences as well as the words distribution in the training data. Analyzing the lengths of sentences, word distribution, and character distribution in the corpus acts as the fundamental way of understanding the text data for both supervised and unsupervised learning

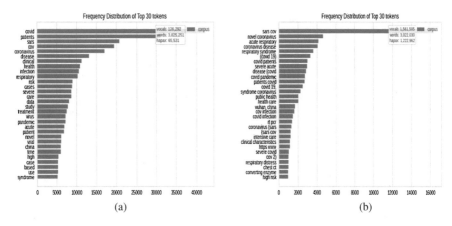

Fig. 2.24 N-grams frequency analysis. (**a**) Top 30 1-grams. (**b**) Top 30 2-grams

from explainability and diagnostic perspective. Figure 2.23a shows data exploration
at a word level, specifically histogram plot showing the number of words appearing
in each document. Figure 2.23b shows distribution of average word length in each
sentence.

2.2.7.2   N-Grams Analysis

N-grams analysis involves studying the most frequent n-grams used in the corpus
either in a supervised or in an unsupervised manner. This analysis helps to understand better the context in which the word gets used in the NLP task. These n-grams
are generally the features extracted from the text for traditional machine learning
models, and hence understanding the dominant phrases becomes imperative from
an explainability outlook. In the real world, frequency analysis of one to three n-
grams at the word level is the most common practice. Figure 2.24a and b shows
1-gram and 2-gram analysis for the LitCovid dataset.

2.2.7.3   Word Cloud

Word cloud is a technique for visualizing the most frequent words in a corpus where
the color and size of the words in the image represents their frequency. The word
cloud can be used for the entire corpus or in the labeled dataset for a particular class
to visualize the words that dominate. Word cloud is also used as a diagnostic tool
to identify and eliminate corpus specific stop words, decision for lemmatization,
handling of diacritics, etc. Figure 2.25 shows the word cloud for the LitCovid
dataset.

Fig. 2.25 Word cloud for the LitCovid dataset

2.2.7.4 Topic Modeling

Topic models are unsupervised statistical language models used for revealing the latent structures and categories in the corpus. The basis of topic modeling is that each document has a probability distribution over topics, and each topic is associated with a distribution over terms. Topic modeling methods such as Latent Semantic Analysis (LSA), Probabilistic Latent Semantic Analysis (PLSA), Non-negative Matrix Factorization (NMF), and Latent Dirichlet Analysis (LDA) are some of the well-known techniques used to find the topics and term distributions in the corpus [AA15]. Latent Dirichlet Allocation is one of the most practical and effective traditional methods for topic modeling. Feature extraction and selection can also use topic modeling for identifying important terms, categories, and even cleaning the corpus of irrelevant data in a large text corpus. Figure 2.26 shows topics and word distributions using LDA.

2.2.7.5 Corpus Visualization

Corpus visualization using dimensionality reduction and visualization techniques such as Principal Component Analysis (PCA), t-distributed Stochastic Neighbor Embedding(t-SNE), and Uniform Manifold Approximation and Projection (UMAP) are well suited to visualize the documents with labels as a scatter plot. Some of these techniques are very effective in visualizing the clusters along with relative proximities and overlap between them.

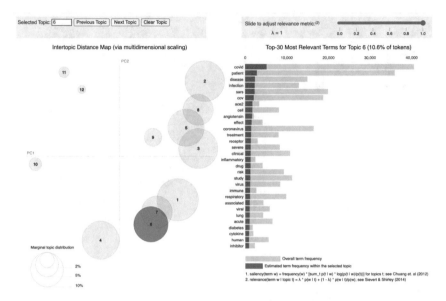

Fig. 2.26 Topic modeling on LitCovid dataset using LDA and pyLDA visualization

```
Topic 0:
health care pandemic social public people healthcare services medical crisis research global community many impact
Topic 1:
cov sars protein coronavirus virus human viral respiratory binding rna spike coronaviruses mers syndrome host
Topic 2:
patients respiratory acute severe disease coronavirus clinical syndrome patient cardiac care infection symptoms risk failure
Topic 3:
il cells immune inflammatory cytokine cell anti lung cytokines inflammation storm response induced activation viral
Topic 4:
cases china epidemic coronavirus number transmission outbreak novel confirmed countries case wuhan control january measures
Topic 5:
p patients ci , (0 study 95 data ( vs (n analysis age group 001
Topic 6:
hydroxychloroquine treatment chloroquine drug drugs hcq trials clinical antiviral azithromycin trial use mg remdesivir anti
Topic 7:
ct chest pneumonia imaging lung findings glass ground patients opacities bilateral pulmonary consolidation patient rt
Topic 8:
surgical surgery procedures patients staff ppe masks mask equipment patient healthcare protective aerosol risk room
Topic 9:
cancer patients breast treatment surgery oncology pandemic radiation radiotherapy chemotherapy risk therapy care colorectal management
Topic 10:
ace2 angiotensin receptor enzyme converting expression ace ii inhibitors ang renin ras lung blockers cells
Topic 11:
pcr cov sars testing positive rt igg detection negative samples test igm assay tests antibody
```

Fig. 2.27 Topic modeling on LitCovid dataset using non-negative matrix factorization

Observations:

- Figure 2.23a shows that most abstracts are around 250-300 words and the range is as high as around 600 words for some documents.
- Figure 2.23b shows that most sentences are of length 6, but there are some long sentences with 18 words.

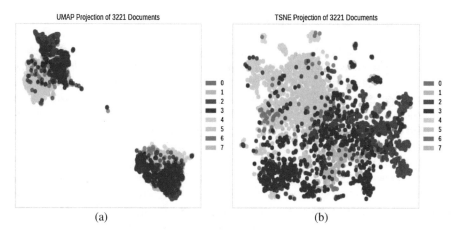

Fig. 2.28 Corpus visualization. (**a**) UMAP. (**b**) t-SNE

- Figure 2.25 which highlights the most frequent words in the corpus not only captures the theme or the topics around Covid-19 but also indicates the need for preprocessing like lemmatization. For example, the noun **virus** and its plural form **viruses** appear in the top hits and needs handling.
- Figure 2.24a shows top words in the corpus, highlighting the words such as **coronavirus**, **cov**, and **covid** may be all synonyms and need domain-specific preprocessing for modeling.
- Figures 2.26 and 2.27 show topic modeling results using LDA and NMF, respectively. The NMF results are very interesting as it surfaces many latent topics, for example, **Topic 10** as **comorbidities and procedures** with words such as cancer, breast, oncology, radiation, radiotherapy, chemotherapy, etc., similarly, **Topic 4** as **origin** with words such as China, confirmed, Wuhan, January, etc.
- Figure 2.28b shows corpus that visualization with t-SNE is better than UMAP and shows separation between clusters with class labels 2, 4, and 7, i.e., **Forecasting**, **Mechanism**, and **Treatment**.

2.2.8 EDA and Computer Vision

This section will cover techniques that are specific to computer vision, especially image classification.

2.2.8.1   Distributional Analysis

In image recognition and image classification, analyzing the data from the spatial average pixel intensity and aggregate distributional differences help understand how the classes relate. We can show how dark or bright images are by averaging the 28×28 pixel values over each image. Values closer to 1 will have higher intensity, i.e., it has some objects filling in the space, while values closer to 0 will have fewer pixels shaded (or uniformly low-intensity pixels). Figures 2.29 and 2.30 show distribution per class through histograms and boxplots.

By performing an average of pixel values, we can see a "prototypical" version of each category in the training dataset. The analysis helps one to understand if types or classes are close to each other spatially and need more discriminating examples or changes in modeling techniques to categorize them. Figure 2.31 shows images per class for the training data in the Fashion MNIST dataset. We can also approximate how symmetric an image is by taking the absolute value of the difference between a horizontally flipped image and the original (Fig. 2.32). For more asymmetric images, there will be more highlighted pixels in the difference, so averaging the absolute difference will give us a rough quantification of the symmetry of each image.

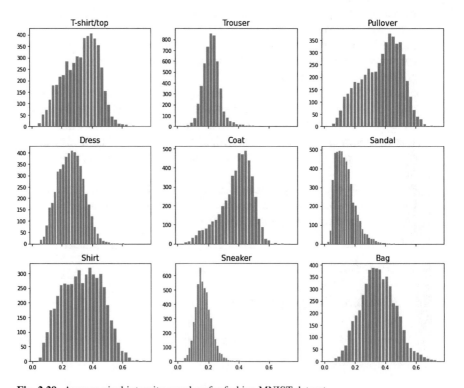

Fig. 2.29 Average pixel intensity per class for fashion MNIST dataset

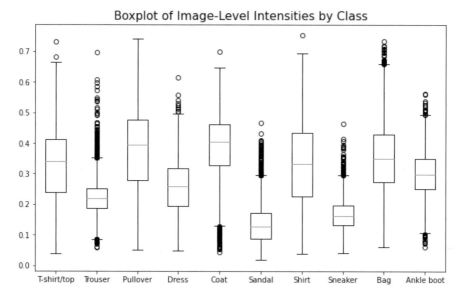

Fig. 2.30 Boxplot of pixel intensity per class for fashion MNIST dataset

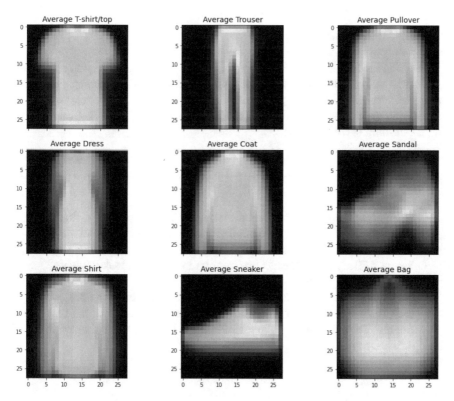

Fig. 2.31 Average pixel intensity per class to understand the prototypes in fashion MNIST dataset

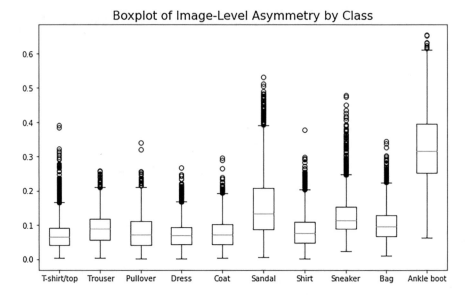

Fig. 2.32 Boxplot capturing symmetry distribution for fashion MNIST dataset

2.2.8.2   2D Projections

Visualizing the image vectors in two-dimensional space using projection techniques such as Principal Component Analysis (PCA) and/or manifold learning techniques such as Uniform Manifold Approximation and Projection (UMAP) helps us see the spread and the overlaps. Figures 2.33 and 2.34 show the plots for PCA and UMAP on the Fashion MNIST dataset.

Observations:

- In Fig. 2.29, we can see that sandals and sneakers tend to skew darker, while pullovers and t-shirt/top have similarly shaped lighter distributions. Shirts have an interesting distribution as well.
- Figure 2.30 highlights the fact that trousers, sneakers, and sandals have many outliers and a long tail.
- In Fig. 2.31, we can identify some features that make up each class: we can see the sleeves of a coat, the trouser legs, and the handle of the bag. We can also see that there is quite a bit of variety in sandals and that shirts and dresses may or may not have sleeves.

- Comparing the linear PCA and the non-linear UMAP in Figs. 2.33 and 2.34, we can say that non-linear dimensionality reduction isolates certain classes better, such as trousers and bags, and aligns with intuition since these classes are each entirely dissimilar from the other categories. UMAP also highlights the relative similarities between specific clusters: the group containing footwear (sneakers, sandals, and ankle boots) is close together, with each class somewhat contained into its subgroup (i.e., relatively little overlap between groups). On the other hand, another cluster includes tops, shirts, coats, pullovers, and dresses; the intra-cluster spread and overlap between classes could indicate some degree of difficulty for our classification task.

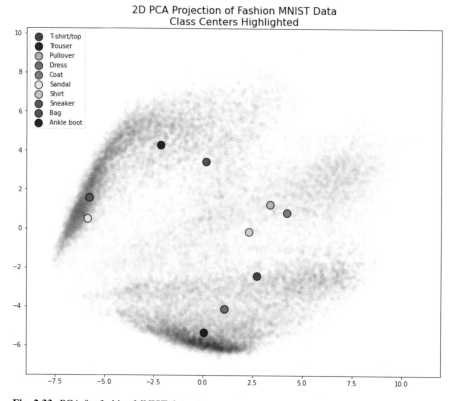

Fig. 2.33 PCA for fashion MNIST dataset

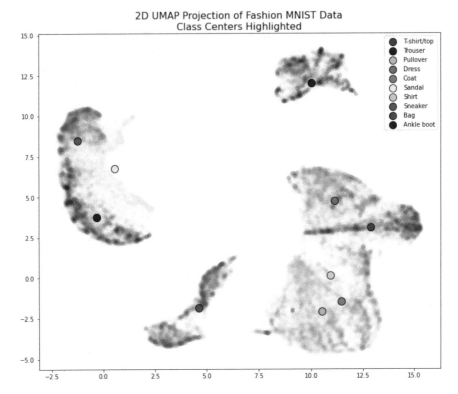

Fig. 2.34 UMAP for fashion MNIST dataset

2.3   Feature Engineering

Feature engineering is a generic term used in machine learning to define multiple
processes such as feature construction, feature selection, feature importance, dimensionality reduction, and explanations [DL18]. As the number of features increases,
feature selection and reduction become critical for the model's performance, like
document or text classification and other NLP tasks. Selecting a subset of k features
typically involves searching, evaluating, and even building models with $2^k - 1$ sets.
These features influence the machine learning models and play a massive role in
explainability. This section will discuss the traditional feature selection methods
that influence interpretability and explainability of the models mostly in a model-
agnostic manner. Model-specific feature selection is described in the next two
chapters.

2.3.1   Feature Engineering and Explainability

How does traditional feature engineering help in explainability?

1. It helps in reducing complexity of the models and hence increasing the interpretability. Any feature selection or reduction technique that removes features helps make the model simpler because of fewer interactions, removing noise caused by inter-relationships, and better predictability of the output.
2. It generally helps make models cheaper with respect to storage, faster in computation, and facilitates better interpretable models. The automated model building across parameters, hyperparameters, and model choices becomes more manageable with smaller feature sets.
3. One of the critical aspects of feature engineering is that it helps in reducing overfitting. Overfitting is the prime cause of the model's poor performance on unseen data, and feature analysis plays an important role here. Understanding the distribution difference between the features present in the training data and the prediction data explains poor performance of many models.

2.3.2   Feature Engineering Taxonomy and Tools

There are many ways to select a subset of features or rank them based on their importance. We will enumerate some of the most practical categories and techniques here.

> Python packages **mlxtend** and **scikit-feature** along with **scikit-learn** are used for our analysis and have comprehensive coverage for most filter, wrapper, search, unsupervised, and embedded methods.

2.3.2.1   Filter-Based

The general approach taken here is to employ any search-based technique and evaluate the importance of the feature(s) using some statistical measure until it reaches a stopping criterion (based on some heuristic). It can be fundamental, like removing constant (zero variance) features or features having variance above a user-defined threshold. Information theoretic techniques such as information gain measure the increase in entropy due to the feature's presence and hence its importance as a score. Statistical techniques using F-score-based methods such as selecting k-best features with high F-scores are common. Statistical methods such as the chi-squared test measure the dependence between the categorical feature and the categorical label. ANOVA is often used in the case of continuous features.

Similarity-Based

Similarity-based techniques work on an underlying principle that the data from the same class are closer in high-dimensional space.

Given a data matrix $\mathbf{X} \in \mathbb{R}^{n \times m}$, with n instances and m features, the features are given by $f = f_1, \ldots, f_m$ and feature vectors are $\mathbf{f} \in \mathbb{R}^n$. The general framework of similarity-based techniques is to find $\mathbf{S} \in \mathbb{R}^{n \times m}$ such that

$$\max_{S} U(S) = max_S \sum_{f \in S} U(f) = max_S \sum_{f \in S} \hat{\mathbf{f}}' \hat{\mathbf{S}} \hat{\mathbf{f}} \tag{2.26}$$

where $U(S)$ is the utility of the feature subset, $U(f)$ is the utility of the feature, $\hat{\mathbf{f}}$ is the transformation of feature vector, and $\hat{\mathbf{S}}$ is the transformation of similarity matrix. Based on the transformations $\hat{\mathbf{f}}$ and $\hat{\mathbf{S}}$, there are different selection algorithms.

He et al. proposed Laplacian Score technique to select features such that they maintain sample locality specified by the similarity matrix $\mathbf{S}$. Given $\mathbf{S}$, its corresponding diagonal matrix $\mathbf{D}$, and Laplacian matrix $\mathbf{L}$, the Laplacian score is given by

$$score_{LS}(f_i) = \frac{\tilde{\mathbf{f}}_i^{\mathrm{T}} \mathbf{L} \tilde{\mathbf{f}}_i}{\tilde{\mathbf{f}}_i^{\mathrm{T}} \mathbf{D} \tilde{\mathbf{f}}_i}, \ where \ \tilde{\mathbf{f}}_i = \mathbf{f}_i - \frac{\mathbf{f}_i^{\mathrm{T}} \mathbf{D} \mathbf{1}}{\mathbf{1}^{\mathrm{T}} \mathbf{D} \mathbf{1}} \tag{2.27}$$

The score has a numerator that captures the consistency of features on the similarity matrix; the smaller, the better, and denominator feature variance, the higher, the better. Thus, the smaller the score, the better it is for the selected feature.

Zhau and Liu's Spectral Feature Selection is an extension of Laplacian Score technique, where eigenvectors of similarity matrix are used to represent the data distribution with the assumption that eigenvectors of similar data are of the same affiliations. Higher scores are preferred in the technique.

$$score_{SPEC}(f_i) = \sum_{j=1}^{n} \lambda_j \left(\varepsilon_j^{\mathrm{T}} \mathbf{f}_i \right) = \mathbf{f}_i^{\mathrm{T}} \mathbf{S} \mathbf{f}_i \tag{2.28}$$

Given the class labels $y = y_1, \ldots, y_n$, Duda et al. proposed using the Fischer score for selecting features within the class ($\mathbf{S}^w$) and between classes ($\mathbf{S}^b$) by defining the local and global similarity matrices, respectively, as:

$$\mathbf{S}_{i,j}^w = \begin{cases} 1/n_l \ if \ y_i = y_j = l \\ 0 \ otherwise \end{cases} \tag{2.29}$$

$$\mathbf{S}_{i,j}^b = \begin{cases} 1/n - 1/n_l \ if \ y_i = y_j = l \\ 1/n \ otherwise \end{cases} \tag{2.30}$$

If $\mathbf{L}^w$ and $\mathbf{L}^b$ are the Laplacian matrices from $\mathbf{S}^w$ and $\mathbf{S}^b$, respectively, the Fischer score is given by

$$score_{FS}(f_i) = \frac{\tilde{\mathbf{f}}_i^{\mathrm{T}}\mathbf{L}^b\tilde{\mathbf{f}}_i}{\tilde{\mathbf{f}}_i^{\mathrm{T}}\mathbf{L}^w\tilde{\mathbf{f}}_i} \tag{2.31}$$

The larger the Fischer score, the better is the selected feature.

Relief (and its multiclass extension ReliefF) weighs features based on the difference between distance to nearest data points with the same class label and distance to nearest data points with different class labels. It is given by:

$$score_R(f_i) = \frac{1}{2}\sum_{t=1}^{p} d(f_{t,i} - f_{NM(\mathbf{x}_t),i}) - d(f_{t,i} - f_{NH(\mathbf{x}_t),i}) \tag{2.32}$$

where $f_{t,i}$ is the value of instance $\mathbf{x}_t$ for the feature $f_i, f_{NM(\mathbf{x}_t),i}, f_{NH(\mathbf{x}_t),i}$ denote the values on the i-th feature of the nearest points to $\mathbf{x}_t$ with the same and different class labels, respectively, and $d(\cdot)$ is a distance metric.

Similarity-based feature selection techniques are simple and easy to calculate, and the selected features help in subsequent learning tasks. One of the disadvantages is that most techniques do not handle feature redundancy.

Figure 2.35 compares the Fischer score and ReliefF for the diabetes dataset.

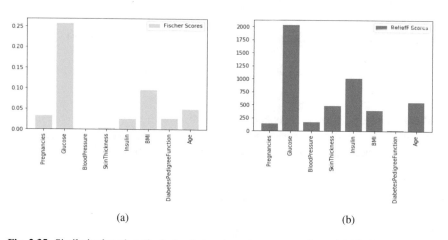

(a)   (b)

Fig. 2.35 Similarity-based methods for feature importance comparison. (**a**) Fischer scores. (**b**) ReliefF scores

Information Theoretic-Based

Feature selection by maximizing the mutual information between the selected features and the label forms the basis for many information theoretic approaches. Information gain between X and Y is given by

$$I(X; Y) = H(X) - H(Y) = \sum_{x_i \in X} \sum_{y_j \in Y} P(x_i, y_j) log\left(\frac{P(x_i, y_j)}{P(x_i)P(y_j)}\right) \qquad (2.33)$$

Similarly, conditional information gain is given by

$$I(X; Y|Z) = H(X|Z) - H(X|Y, Z)$$

$$= \sum_{z_k \in Z} P(z_k) \sum_{x_i \in X} \sum_{y_j \in Y} P(x_i, y_j|z_k) log\left(\frac{P(x_i, y_j|z_k)}{P(x_i|z_k)P(y_j|z_k)}\right)$$

$$(2.34)$$

Searching for the best feature subset is an NP-hard problem, and most methods employ forward/backward sequential search heuristics to find the best subset. The general goal is to maximize correlation between feature f_i and the class label Y given by $I(f_i; Y)$, minimize the redundancy between the selected features in the subset S given by $\sum_{f_j \in S} I(f_j; f_k)$, and maximize the complementary information with respect to selected features and given by $\sum_{f_j \in S} I(f_j; f_k|Y)$. The generalized scoring can thus be written as

$$score(f_k) = I(f_k; Y) + \sum_{f_j \in S} g[I(f_j; f_k), I(f_j; f_k|Y)] \qquad (2.35)$$

where $g(\cdot)$ is generally a linear function and can be further decomposed as

$$score(f_k) = I(f_k; Y) - \beta \sum_{f_j \in S} (f_j; f_k) + \lambda \sum_{f_j \in S} I(f_j; f_k|Y)] \qquad (2.36)$$

where β and λ have values between 0 and 1.

Information gain only measures the feature importance by its correlation with class labels and can be considered a special case of the above with $\beta = \lambda = 0$ and is given by

$$score_{IG}(f_k) = I(f_k; Y) \qquad (2.37)$$

Mutual information overcomes the shortcoming by considering the redundancy in features too as given by

Fig. 2.36 Information theoretic-based mutual information for feature importance

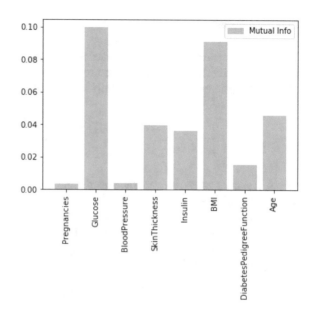

$$score_{MI}(f_k) = I(f_k; Y) - \beta \sum_{f_j \in S} (f_j; f_k) \tag{2.38}$$

and can be considered as a special case with $\lambda = 0$ (Fig. 2.36).

Peng et al. devised Minimum Redundancy Maximum Relevance (mRMR), which is accomplished by adjusting the β in the Information gain by the norm of the similarity matrix as given below:

$$score_{mRMR}(f_k) = I(f_k; Y) - \frac{1}{|S|} \sum_{f_j \in S} (f_j; f_k) \tag{2.39}$$

The information theoretic-based feature selection techniques in addition to carrying all the advantages of similarity-based methods also handle both feature relevance and redundancy. Information theoretic techniques work mostly on supervised data and need discretization for continuous features. They can capture any statistical dependency, but being nonparametric requires more data samples.

Statistical-Based

Statistical measures can be used to measure the importance of the features. t-Test compares the means between the two groups to see if the feature makes the means of samples from two classes statistically significant.

$$score_{ttest}(f_i) = \frac{|(\mu_1 - \mu_2)|}{\sqrt{\frac{\sigma_1^2}{n_1} + \frac{\sigma_2^2}{n_2}}}$$ (2.40)

where μ_1, μ_2 and σ_1, σ_2 are the means and standard deviations of two binary classes. The higher the t-test score, the more important the feature is.

ANOVA or F-value scoring can be used for selecting importance of continuous features with respect to the target or the label by computing the sum of squares between the group (SSG) and the sum of squares within the group (SSE) and their respective degrees of freedom as given below:

$$score_F = \frac{SSE/df_G}{SSG/df_E}$$ (2.41)

Chi-squared test uses independence test to assess whether the feature is independent of the target and scores the relative importance of the feature based on that. Given a feature with r values

$$score_{chi}(f_i) = \sum_{j=1}^{r} \sum_{s=1}^{c} \frac{(n_{js} - \mu_{js})^2}{\mu_{js}} \quad where \ \mu_{js} = \frac{n_{*s} n_{j*}}{n}$$ (2.42)

n_{js} are instances with j-th feature value in class c, n_{*s} is the number of instances of class c, and n_{j*} is the number of instances with j-th feature value.

Figure 2.37 shows two statistical methods used for feature importance on the diabetes dataset.

The F-value methods such as ANOVA estimate the degree of linear dependency between two random variables and have assumptions about the feature's distribution. Many statistical measures are not very effective in high-dimensional data. Most measures cannot handle redundancy.

Filter-based techniques have the advantage of being model-agnostic and relatively faster than the wrapper-based approach.

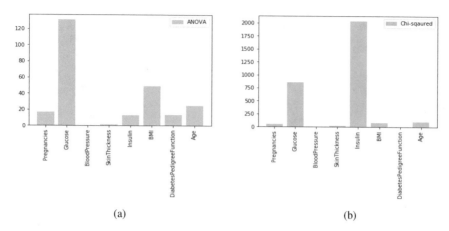

Fig. 2.37 Statistical methods-based feature importance comparison. (**a**) ANOVA F-statistic scores. (**b**) Chi-squared statistics score

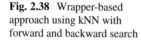

Fig. 2.38 Wrapper-based approach using kNN with forward and backward search

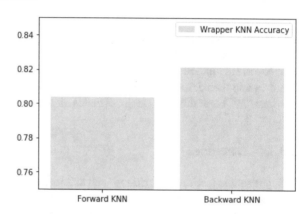

2.3.2.2   Wrapper-Based

In contrast to the filter-based techniques, the wrapper-based approach uses one or more machine learning algorithms to evaluate features. The wrapper approach can use the same algorithm that the modeling step uses or can use a completely different algorithm based on speed, performance, and interpretability. Since the features are evaluated based on the model performance, it is imperative to select the right metrics such as F-score, precision, recall, accuracy, etc. for the model and therefore the feature selection.

2.3.2.3   Unsupervised

As discussed in the multivariate visualization, unsupervised linear techniques such as PCA, MDS, t-SNE, Isomap etc. can be used as feature reduction, selection,

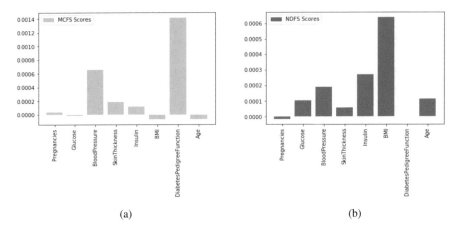

Fig. 2.39 Unsupervised techniques for feature importance comparison. (**a**) MCFS for feature importance. (**b**) NDFS for feature importance

and feature generation techniques. Some of the methods such as Multi-Cluster Feature Selection (MCFS), Nonnegative Discriminative Feature Selection (NDFS), and Norm Regularized Discriminative Feature Selection can be used for analyzing features and their importance [CZH10, Li+12, Yan+11]. Figure 2.39 compares and contrasts MCFS and NDFS on the diabetes dataset.

2.3.2.4   Embedded

Many machine learning algorithms such as Linear Regression, Logistic Regression, Decision Trees, Random Forest, Boosting, and many others have feature selection as a component in the modeling. Regularization techniques such as L1 and L2 also play the role of feature selection in many of these algorithms. This will be discussed in detail in the next chapter.

Observations:

- The two statistical-based feature selection algorithms ANOVA and chi-squared are shown in Fig. 2.37a and b, respectively. ANOVA ranks *Glucose*, *BMI*, and *Age* as the top three features, while chi-squared ranks *Insulin*, *Glucose*, and *Age* at the top. *Pregnancies* and *BloodPressure* both show relatively low scores in both the techniques.
- The information theoretic-based mutual information method ranks *Glucose*, *BMI*, and *Age* as the top three features similar to ANOVA.

- Both the similarity-based methods, as seen in Fig. 2.35, show different high scores for many features. The Fischer score suggests *Glucose*, *BMI*, and *Age* as the top feature, while ReliefF shows *Glucose*, *Insulin*, and *Age* as the most discriminating.
- The wrapper-based feature selection as shown in Fig. 2.38 gives different accuracies with the same classifier (K-nearest neighbors) by choosing a different combination of features. The forward search results in feature set of *Pregnancies, Glucose, BloodPressure, BMI*, and *DiabetesPedigreeFunction* with a classification accuracy of 80.35. The backward search shows features *Pregnancies,Glucose, SkinThickness, BMI*, and *Age* as the chosen set resulting in a higher classification accuracy of 82.1.
- The unsupervised feature selection methods MCFS and NDFS show completely different scores for the features as shown in Fig. 2.39. The MCFS method shows *DiabetesPedigreeFunction* and *BloodPressure* as the top feature, while NDFS indicates *BMI* and *Insulin* as the most important features.

References

[Agg16] C.C. Aggarwal, *Outlier Analysis*. 2nd edn. (Springer Publishing Company, Incorporated, Berlin, 2016). ISBN: 3319475770

[AA15] R. Alghamdi, K. Alfalqi, A survey of topic modeling in text mining. Int. J. Adv. Comput. Sci. Appl. **6**(1) (2015). https://doi.org/10.14569/IJACSA.2015.060121

[And94] T.W. Anderson, *An Introduction to Multivariate Statistical Analysis* (Wiley, Hoboken, 1994)

[Ber19] F. Bertrand, *SweetViz: Exploratory Data Analysis for Python* https://github.com/fbdesignpro/sweetviz.2019

[Bik19] N. Bikakis, Big data visualization tools, in *Encyclopedia of Big Data Technologies*, ed. by S. Sakr, A.Y. Zomaya (Springer International Publishing, Cham, 2019), pp. 336–340. ISBN: 978-3-319-77525-8. https://doi.org/10.1007/978-3-319-77525-8_109

[Bok20] Bokeh Development Team. *Bokeh: Python Library for Interactive Visualization* (2020). https://bokeh.org/

[BD91] P.J. Brockwell, R.A. Davis, *Time Series: Theory and Methods*, 2nd edn. (Springer, Berlin, 1991)

[BD96] P.J. Brockwell, R.A. Davis, *Introduction to Time Series and Forecasting* (Springer, New York, 1996)

[Bru19] S. Brugman, *Pandas-Profiling: Exploratory Data Analysis for Python* (2019) https://github.com/pandas-profiling/pandas-profiling

[CZH10] D. Cai, C. Zhang, X. He, Unsupervised feature selection for multi-cluster data, in *Proceedings of the 16th ACM SIGKDD International Conference on Knowledge Discovery and Data Mining*. KDD '10 (Association for Computing Machinery, Washington, 2010), pp. 333–342. ISBN: 9781450300551. https://doi.org/10.1145/1835804.1835848

[CBK09] V. Chandola, A. Banerjee, V. Kumar, Anomaly detection: a survey. ACM Comput. Surv. **41**(3), 15:1–15:58 (2009). https://doi.org/10.1145/1541880.1541882

[Cri20] E. De Cristofaro, *An Overview of Privacy in Machine Learning* (2020). arXiv: `2005.08679 [cs.LG]`

[DL18] G. Dong, H. Liu, *Feature Engineering for Machine Learning and Data Analytics*, 1st edn. (CRC Press, Boca Raton, 2018). ISBN: 1-13874-438-7

[Eds03] R.M. Edsall, The parallel coordinate plot in action: design and use for geographic visualization. Comput. Stat. Data Anal. **43**(4), 605–619 (2003). https://doi.org/10. 1016/S0167-9473(02)00295-5

[Gab71] K.R. Gabriel, The biplot graphical display of matrices with applications to principal component analysis. Biometrika **58** , 453–467 (1971)

[Gho+18] A. Ghosh, et al. A comprehensive review of tools for exploratory analysis of tabular industrial datasets. Vis. Inf. **2**(4), 235–253 (2018) https://doi.org/10.1016/j.visinf. 2018.12.004

[Gra09] J.W. Graham, Missing data analysis: making it work in the real world. Ann. Rev. Psychol. **60**, 549–576 (2009)

[Gri18] B. Griner, *Decoding Health with Data Science and Machine Learning* (2018). https:// briangriner.github.io/decoding-health-risk-factors-pre-diabetes-ML-3.5.18.html

[Hab72] S.J. Haberman, Log-linear fit for contingency tables—Algorithm AS51. Appl. Statist. **21**, 218–225 (1972)

[HM82] A.C. Harvey, C.R. McKenzie, Algorithm AS182. An algorithm for finite sample prediction from ARIMA processes. Appl. Statist. **31**, 180–187 (1982)

[Hun07] J.D. Hunter, Matplotlib: a 2D graphics environment. Comput. Sci. Eng. **9**(3), 90–95 ((2007)). https://doi.org/10.1109/MCSE.2007.55

[HA18b] R.J. Hyndman, G. Athanasopoulos, *Forecasting: Principles and Practice*, English. 2nd edn. (OTexts, Melbourne, 2018)

[Inc15] Plotly Technologies Inc. *Collaborative Data Science* (2015). https://plot.ly

[JT09] I.M. Johnstone, D.M. Titterington, Statistical challenges of high dimensional data. Philosoph. Trans. Roy. Soc. London Ser. A **367**(1906), 4237–4253 (2009). https:// doi.org/10.1098/rsta.2009.0159

[Jon80] R.H. Jones, Maximum likelihood fitting of ARMA models to time series with missing observations. Technometrics **22**, 389– 395 (1980)

[KLW19a] U. Kamath, J. Liu, J. Whitaker, *Deep Learning for NLP and Speech Recognition* (Springer, Berlin, 2019). ISBN: 978-3-030-14595-8. https://doi.org/10.1007/978-3- 030-14596-5

[Li+12] Z. Li, et al., Unsupervised feature selection using nonnegative spectral analysis, in *Proceedings of the National Conference on Artificial Intelligence*, vol. 2 (2012), pp. 1026–1032

[MH08] L. van der Maaten, G. Hinton, Visualizing data using t-SNE. J. Mach. Learn. Res. **9**, 2579–2605 (2008). http://www.jmlr.org/papers/v9/vandermaaten08a.html

[MKB79] K.V. Mardia, J.T. Kent, J.M. Bibby, *Multivariate Analysis* (Academic, London, 1979)

[Scho+2018] P. Schober, C. Boer, L.A. Schwarte, Correlation coefficients: appropriate use and interpretation, *in Anesthesia & Analgesia*, vol. 126, (Wolters Kluwer, 2018), pp. 1763–1768

[SP10] S. Seabold, J. Perktold, Statsmodels: Econometric and statistical modeling with python, in *9th Python in Science Conference* (2010)

[Sel] H.J. Seltman, *Experimental Design and Analysis, EPUB* (Carnegie Mellon University, Pittsburgh, PA, 2018). http://www.stat.cmu.edu/~hseltman/309/Book/Book.pdf

[SGM10] F. Song, Z. Guo, D. Mei, Feature selection using principal component analysis, in *2010 International Conference on System Science Engineering Design and Manufacturing Informatization*, vol. 1 (2010), pp. 27–30

[Ste+19] J.-L. Stevens, et al., *pyviz/holoviews: Version 1.12.5*. Version v1.12.5 (2019). https://doi.org/10.5281/zenodo.3368625

[tea20] The pandas development team. *pandas-dev/pandas: Pandas* Version lat- est. (2020). https://doi.org/10.5281/zenodo.3509134

[TSL00] J.B. Tenenbaum, V. de Silva, J.C. Langford, A global geometric framework for nonlinear dimensionality reduction. Science **290**(5500), 2319–2323 (2000). ISBN: 0036-8075. https://science.sciencemag.org/content/29/550/2319.full.pdf. https://science.sciencemag.org/content/290/5500/2319

[Tuf11] S. Tuffery, *Data Mining and Statistics for Decision Making*. Wiley Series in Computational Statistics (Wiley, Hoboken, 2011). ISBN: 9780470979280. https://books.google.com/books?id=5MTBlxZUKiIC

[Tuk77] J.W. Tukey, *Exploratory Data Analysis* (Addison-Wesley, Reading, 1977)

[Wt20] Michael Waskom and the seaborn development team. *mwaskom/seaborn* Version latest. (2020). https://doi.org/10.5281/zenodo.592845

[Yan+11] Y. Yang, et al., L2,1-norm regularized discriminative feature selection for unsupervised learning, in *Proceedings of the Twenty-Second International Joint Conference on Artificial Intelligence Volume Two*. IJCAI'11 (AAAI Press, Barcelona, 2011), pp. 1589–1594. ISBN: 9781577355144

Chapter 3
Model Visualization Techniques and Traditional Interpretable Algorithms

One of the easiest ways to build explainable models is by having the machine learning algorithm be intrinsically interpretable. Gaining an understanding of how well a model performs from looking at the results of model evaluation is another important way to enhance model explainability. We discuss several techniques to visualize model evaluation including precision-recall curves, ROC curves, residual plots, silhouette coefficients, and others to give a comprehensive overview of classification, regression, and clustering techniques. Next, we start understanding interpretability of some of the traditional machine learning models used in classification, regression, and clustering. The Pima Indian diabetes dataset is used to perform supervised and unsupervised classification. The insurance claims dataset is used for regression model analysis.

3.1 Model Validation, Evaluation, and Hyperparameters

The key to creating great models is to make sure that the model generalizes well on unseen data. Figure 3.1 gives the most well-established process that ensures models do not overfit (or underfit) and generalize well for classification and regression [HTF09a]. The labeled dataset can be divided into training, validation, and test sets from the original data. Primarily, the test set should be representative of the unseen real-world data in terms of quality, distribution, class balance, etc. If it is representative, running the model and evaluating the metrics on the test data gives an estimate close to what real-world model performance will be. Most algorithms have various parameters or options that have to be set for optimal performance. Generally, a separate validation set is used for evaluating model performance on different parameter values. In the absence of a separate validation set, splitting training data into train and validation sets is a choice and depends on the amount of labeled data and the model capacity (VC dimensions). Validation techniques like k-fold cross-

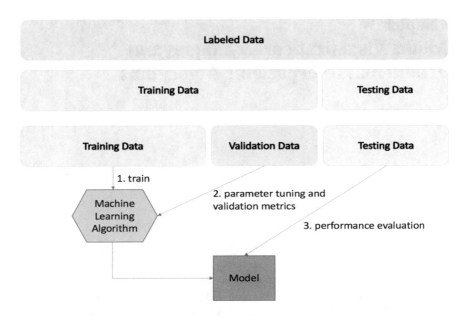

Fig. 3.1 Training, validation, and test sets for model tuning and evaluation

validation are employed when separate validation sets are not a possibility [CT10]. The validation process plays a vital role in tuning or selecting the model parameters. The choice of these parameters affects the model performance, and hence explicitly understanding the options is critical from an explainability standpoint.

> To compare and contrast machine learning models it is necessary to use the same split of train, validation, and test sets to evaluate all the models (with parameters) using the same performance metric(s). Interpretability is also one of the aspects that one should focus on along with other metrics.

3.1.1 Tools and Libraries

For all the tasks related to model performance analysis and visualization of results, we will use the **YellowBrick** package along with **sklearn** on the Pima Indian diabetes dataset (classification) and the insurance claims dataset (regression).

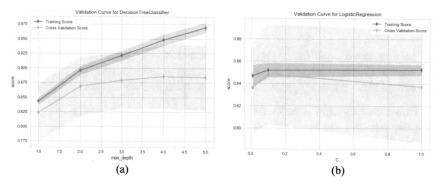

Fig. 3.2 Validation curves for classifiers with AUC—area under ROC curve. (**a**) Decision tree. (**b**) Logistic regression

3.2   Model Selection and Visualization

Most machine learning algorithms have parameters that need to be tuned for optimal performance on a given dataset. For example, a decision tree can have different values of "max depth" and the models corresponding to each such value can exhibit a range of performance values, measured as accuracy or precision, for example. A validation set or cross-validation technique is used to tune these parameters.

3.2.1   Validation Curve

Validation curve is a plot of performance metrics such as a score with respect to different values of the parameters of the model [Bra97].

Observations:

- The validation curve as in Fig. 3.2a for Decision Tree shows that at "max depth" of 4, the classifier stabilizes to give optimum AUC of around 0.88. As the number of nodes increases, the validation score remains almost constant while the training score increases indicating overfitting.
- The validation curve as in Fig. 3.2b for Logistic Regression shows best performance for the parameter C at 0.1 with AUC value around 0.77. As the C value increases the validation score drops indicating the region of overfitting.
- The variance in validation and training scores is very high in Logistic Regression as compared to Decision Tree.

3.2.2 Learning Curve

A learning curve explains the relationship between a performance metric, such as accuracy for a classifier, and the number of training samples [Per10]. The learning curve provides various diagnostic insights into the classifier such as

1. How many training samples does the classifier/regressor need for an optimum performance score in training and validation?
2. Are the samples representative of the domain?
3. Does the bias or the variance introduce error in the classifier/regressor?
4. Does the model have any overfitting or underfitting issues?

The training and validation learning curves are plotted together so we can look at the relative metrics to get the overall diagnosis for decision trees and logistic regression as shown in Fig. 3.3a and b.

- A flat training and validation learning curve indicates a high chance of underfitting as it might signify no improvement and hence no learning.
- A training learning curve indicating a continuous decrease right from the start is also indicative of underfitting.
- High variability in the validation learning curve, especially with cross-validation, but not in the training learning curve indicates error due to variance rather than bias.
- High variability in the training learning curve indicates error due to bias.
- A large gap between the training and validation learning curve diverging after a point in the curve indicates the ideal split and marks the beginning of overfitting.

Observations:

- The learning curves in Fig. 3.3a for Decision Tree show that training and validation curves are separated. At about 600 samples, the validation curve trends downwards. There is a large variance in the cross-validation as compared to training indicating variance errors in predictions rather than bias errors.

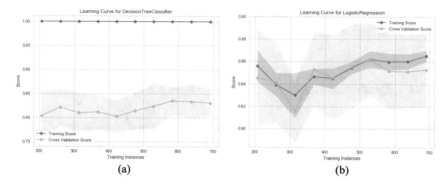

Fig. 3.3 Learning curves for classifiers with AUC—area under ROC curve. (**a**) Decision tree. (**b**) Logistic regression

- The learning curves in Fig. 3.3b for Logistic Regression show both training and validation curves following similar trends and at about 600 samples, showing divergence. Similar to the Decision tree, logistic regression also indicates variance error.
- The training learning curve for Logistic Regression also shows variability and this indicates the bias error. When compared with decision tree, it can be concluded that the non-linear decision tree algorithm performs better indicating the presence of non-linear boundaries.
- The variance in logistic regression is more than that of decision tree.

3.3   Classification Model Visualization

As discussed in the last section, model selection happens based on the agreed metrics that vary based on the domain and the nature of the application [Ras20]. For example, in some compliance-based domains in financial services, false negatives have to be minimized (recall-centric), while in other applications such as fraud detection where there are fewer resources to investigate the positive hits, false positive minimization becomes imperative (precision-centric).

Many model governance teams consider model metrics and evaluation results along with the actual model as an artifact that needs to be documented and reported. From a diagnostic and white-boxing perspective, understanding how the model performs in various scenarios is critical. This section will discuss some well-known model metrics and how they impact selection, especially of the classification models.

3.3.1 Confusion Matrix and Classification Report

As shown in Fig. 3.4, the confusion matrix is a common way to visualize the classification results on the test dataset. It acts both as a quantitative metrics provider for making decisions such as how well the model generalizes and also as a diagnostics tool to understand the model's behavior on individual classes.

Classification report is another view of the confusion matrix but with various metrics that highlight model behavior from an efficiency and effectiveness standpoint. As shown in Fig. 3.5, various metrics such as precision, recall, F1, and support per-class basis are given in the classification report as color-coded heatmaps for Decision Tree model.

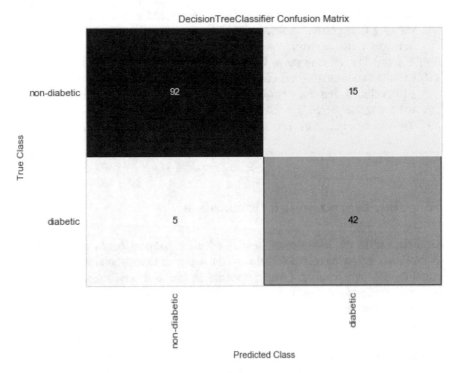

Fig. 3.4 Confusion matrix for decision tree model on diabetes classification dataset

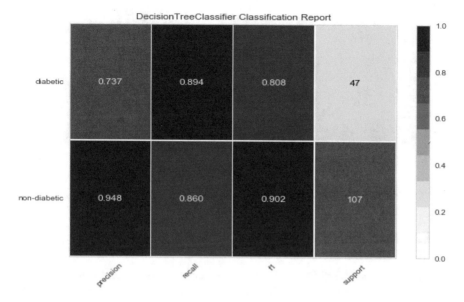

Fig. 3.5 Classification report for decision tree model on diabetes classification dataset

By visualizing classification reports for various models on the same evaluation dataset, model behaviors can be understood in a comparative sense, as shown in Fig. 3.6a and b for Gaussian Naive Bayes and Logistic regression, respectively.

Observations:

- Figure 3.6a shows precision for the diabetic class for the Gaussian Naive Bayes model (68 9) is slightly higher than that of Logistic Regression (68.3). The precision for the non-diabetic class for the Gaussian Naive Bayes model (85.3) is higher than that of Logistic Regression (83.2). Thus if precision is the metric, then Gaussian Naive Bayes is the model one should select.
- Figure 3.6b shows recall for the diabetic class for the Gaussian Naive Bayes model (66) is higher than that of Logistic Regression (59.6). But the recall for the non-diabetic class for Logistic Regression (87.9) is slightly higher than that of Gaussian Naive Bayes (86.9). The choice of the model then depends on the skew of the dataset and the bias towards the predictions of a particular class.
- The F1 score for Gaussian Naive Bayes for both diabetic and non-diabetic is higher than that of Logistic Regression.
- Comparing Figs. 3.5, 3.6a, and b, one can clearly see that for all the metrics such as precision, recall, and F1, the non-linear decision tree model is superior to both Gaussian Naive Bayes and Logistic Regression.

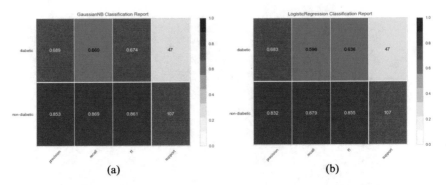

Fig. 3.6 Comparing classification reports for two models. (**a**) GaussianNB. (**b**) Logistic regression

3.3.2   ROC and AUC

The Receiver Operating Characteristic (ROC) curve measures a classifier's predictive quality, comparing and visualizing the trade-off between the model's sensitivity and specificity. Sensitivity measures how often a model correctly generates a positive for the data that is labeled as a positive (also known as the true positive rate). Specificity measures how often a model correctly generates a negative for the data that is labeled as a negative (also known as the true negative rate). The ROC curve generates another metric computing the area under the curve (AUC) and captures the relationship between false positives and true positives [GV18].

> The higher the AUC, the better the model's generalization capability is. The ROC curve's steepness is also crucial as it describes the maximization of the true positive rate while minimizing the false positive rate. The closer the ROC curve is to the top left corner, the better the model's quality is overall. The closer the curve comes to the center diagonal line, the closer the model is to a random guesser.

Observations:

- Figure 3.7a and b show that the AUC for Gaussian Naive Bayes and Decision Tree for both classes are almost identical, with a value of 0.89.
- Based on the steepness of the curve and closeness to the top left corner, Decision Tree seems to be a slightly better choice than Gaussian Naive Bayes

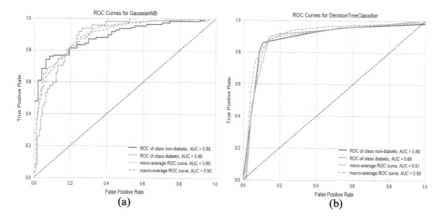

Fig. 3.7 Comparing ROC curves for two models. (**a**) GaussianNB ROC curve. (**b**) Decision tree ROC curve

3.3.3   PRC

Precision-Recall curve measures the trade-off between the two metrics—precision and recall. Precision, measured as a ratio of true positives to the sum of true positives and false positives, is a measure of exactness or efficiency [DG06]. Recall, measured as a ratio of true positives to the sum of true positives and false negatives, is a measure of completeness or effectiveness. Average precision represents the precision-recall curve as a single metric and is computed as the weighted average of precision achieved at each threshold, where the weights are the differences in recall from the previous thresholds.

The larger the area in the Precision-Recall curve, the better is the classifier, especially when there is a huge imbalance between the classes. Higher Average Precision is normally considered a good single metric by which to select the classifier in an imbalanced dataset.

Observations:

- Figure 3.8a and b show the area under PRC for Logistic Regression is higher than that of Gaussian Naive Bayes.
- The average precision for Logistic Regression is more than that of Gaussian Naive Bayes. Hence, in a severely imbalanced dataset, selecting Logistic regression over Gaussian Naive Bayes may seem the right choice.

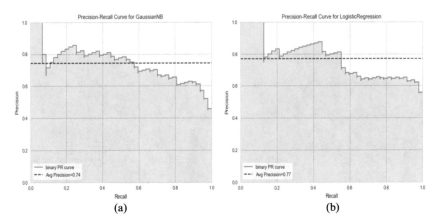

Fig. 3.8 Comparing ROC curves for two classifiers. (**a**) GaussianNB. (**b**) Logistic regression

3.3.4 Discrimination Thresholds

Most classifiers assign a probability for class membership to the instance to be classified. The default is to assume that a probability greater than or equal to 0.5 is for one class and below 0.5 for the other in binary classification. In classification problems with imbalanced data, the default threshold can result in suboptimal performance metrics [Che+05, Pro]. One technique to improve a classifier's performance on imbalanced data is to tune the threshold used to map probabilities to class labels. The discrimination threshold in binary classification, sometimes called classification or decision threshold, is the probability value above which one class is predicted and below which it is the other class.

- Using the training data and creating multiple train/test sets, we run the model multiple times in order to account for the variability in the data. Then the different curves are plotted, showing median and range. The discrimination threshold is the one that achieves the best evaluation metrics in the multiple runs.
- Discrimination threshold tuning is not a hyperparameter tuning but a decision based on the trade-off between false positives and false negatives on the basis of the classifier's probability outputs.
- Tuning the discrimination threshold gives a better trade-off between precision and recall in the precision-recall curves.

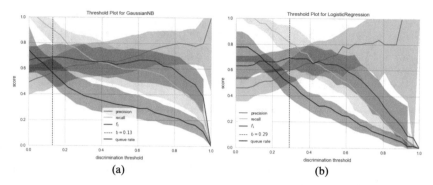

Fig. 3.9 Discrimination thresholds for two classifiers on the diabetes dataset. (**a**) GaussianNB. (**b**) Logistic regression

---

Observations:

- Figure 3.9a and b show the optimal thresholds for Gaussian Naive Bayes and Logistic Regression are 0.29 and 0.13, respectively.
- Gaussian Naive Bayes shows relatively large variance around the mean for the queue rate, F1, and recall while Logistic Regression around precision.

3.4 Regression Model Visualization

Regression model results need to be validated and visualized in a continuous space as compared to classification models. There are various aspects of regression models such as predictions, errors, and sensitivity to hyperparameters that can be used for diagnostics or explainability. In this section, we will discuss some common techniques employed in regression analysis.

3.4.1 Residual Plots

In regression, residual plots plot the difference between the predicted and the observed values for the target. Similar to validation curves and learning curves, residual plots are used for various diagnostics [Bel+80]. The plots can be used to understand the impact of several aspects, for example, outliers, non-linearity of the data, the assumption that the errors are independent and normally distributed and heteroscedasticity.

A good regression residual plot has a high-density of points close to 0 and scattered low density around the axis without a pattern, thus confirming the errors' independence and normal random distribution.

Observations:

- Figure 3.10 shows the residuals for both training and testing data with a good overlap and thus there is no sample bias.
- The errors have multimodal distribution and violate the normal distribution assumption.
- There are patterns around the distribution, especially around +1000 and −1000 value, indicating independence assumption violations.
- The negative spread of errors is more than the positive, showing presence of outliers and long tail.

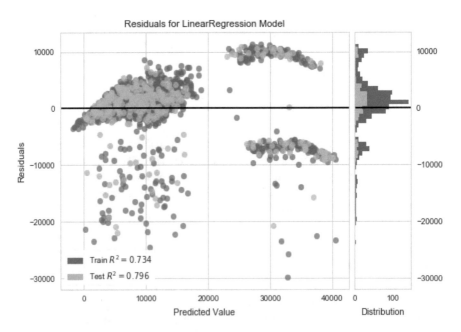

Fig. 3.10 Residual plots for linear regression

3.4.2 Prediction Error Plots

Prediction error plots show the actual values against the predicted values. It also shows the plot with comparison of 45° line.

> Prediction error plots are used for understanding errors caused by variance in the regression model. The comparison with 45° line shows if the model is underestimating or overestimating.

Observations:
Figure 3.11 shows errors are not constant across values, thus variances are not constant and this violates the homoskedasticity assumption.

3.4.3 Alpha Selection Plots

Most regression algorithms employ some form of regularization to constrain the complexity of the model. The alpha values control the complexity of the model and

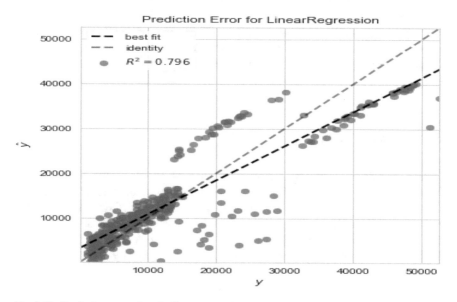

Fig. 3.11 Prediction error plots for linear regression

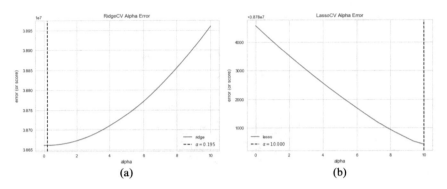

Fig. 3.12 Alpha selection based on errors. (**a**) Ridge regression. (**b**) Lasso regression

cross-validation is used to select the values [HTF09b]. In Fig. 3.12a and b, alpha values that give the lowest error for Ridge Regression and Lasso Regression using cross-validation technique are plotted for the insurance dataset.

> If the alpha values are high, model complexity is reduced, thus reducing the error caused by variance, resulting in an overfit model. If the alpha values are too high, the error due to bias increases, resulting in an underfit model

Observations:
Ridge regression has the lowest error at alpha value of 0.195 and Lasso has lowest error at alpha value of 10.0. Lasso with high alpha values indicates an underfit model with error introduced by the bias.

3.4.4 Cook's Distance

Cook's distance measures an instance's influence on the regression. The larger the influence of an instance, the higher is the likelihood of an outlier, thus influencing the regression model negatively [Coo11]. Visualizing stem plot for all training instances by their Cook's distance score and handling instances with a score more significant than a threshold by removal or imputation is a standard best practice. Cook's distance for ith instance from n observations is given by D_i

$$D_i = \frac{\sum_{j=1}^{n}(\hat{y}_j - \hat{y}_{j(i)})^2}{ps^2} \tag{3.1}$$

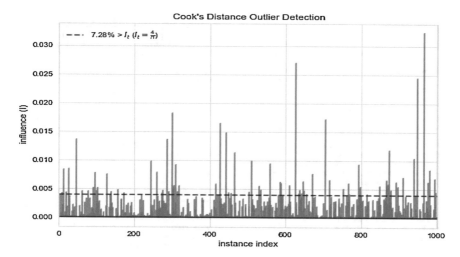

Fig. 3.13 Cook's distance for insurance data

Any instance with score over $4/n$, where n is the number of observations is the threshold for distance scores.

> Any instances with Cook's distance greater than 0.5 or three times the mean score need to be closely examined for their influence.

Figure 3.13 shows the plot of Cook's distance score for the entire insurance training data.

Observations:

There are no instances with distance score greater than 0.5 in the entire dataset. Using just the threshold based on $4/n$, around 7.28% of training data are identified as highly influential based on the Cook's distance scores. Simply removing those instances improves the training R^2 scores from 0.734 to 0.834 as shown in Fig. 3.14a and b, respectively.

3.5 Clustering Model Visualization

Unsupervised learning techniques such as clustering are even more difficult to diagnose or explain as compared to supervised learning since "ground truth" is

OLS Regression Results (a)

Dep. Variable:	y	R-squared:	0.737
Model:	OLS	Adj. R-squared:	0.735
Method:	Least Squares	F-statistic:	371.7
Date:	Wed, 14 Apr 2021	Prob (F-statistic):	1.85e-301
Time:	16:04:43	Log-Likelihood:	-10851.
No. Observations:	1070	AIC:	2.172e+04
Df Residuals:	1061	BIC:	2.177e+04
Df Model:	8		
Covariance Type:	nonrobust		

	coef	std err	t	P>|t|	[0.025	0.975]
age	253.7005	13.530	18.751	0.000	227.152	280.249
bmi	335.9628	32.228	10.424	0.000	272.724	399.201
children	436.9101	156.584	2.790	0.005	129.661	744.159
region_northeast	380.4127	384.691	0.989	0.323	-374.430	1135.255
region_northwest	120.2800	376.809	0.319	0.750	-619.096	859.656
region_southeast	-532.8661	436.403	-1.221	0.222	-1389.177	323.445
region_southwest	-381.5360	391.071	-0.976	0.329	-1148.897	385.825
sex_female	-199.1229	471.328	-0.422	0.673	-1123.964	725.718
sex_male	-214.5866	477.395	-0.449	0.653	-1151.332	722.159
smoker_no	-1.201e+04	478.235	-25.112	0.000	-1.29e+04	-1.11e+04
smoker_yes	1.16e+04	510.457	22.716	0.000	1.06e+04	1.26e+04

Omnibus:	256.825	Durbin-Watson:	1.994
Prob(Omnibus):	0.000	Jarque-Bera (JB):	620.044
Skew:	1.279	Prob(JB):	2.29e-135
Kurtosis:	5.715	Cond. No.	4.39e+17

(a)

OLS Regression Results (b)

Dep. Variable:	y	R-squared:	0.819
Model:	OLS	Adj. R-squared:	0.818
Method:	Least Squares	F-statistic:	574.6
Date:	Wed, 14 Apr 2021	Prob (F-statistic):	0.00
Time:	16:09:03	Log-Likelihood:	-10121.
No. Observations:	1023	AIC:	2.026e+04
Df Residuals:	1014	BIC:	2.030e+04
Df Model:	8		
Covariance Type:	nonrobust		

	coef	std err	t	P>|t|	[0.025	0.975]
age	248.1334	10.815	22.944	0.000	226.912	269.355
bmi	322.2429	26.019	12.385	0.000	271.186	373.299
children	387.6288	125.355	3.092	0.002	141.645	633.613
region_northeast	292.7949	306.909	0.954	0.340	-309.454	895.044
region_northwest	-206.3696	301.218	-0.685	0.493	-797.452	384.712
region_southeast	-564.3746	350.623	-1.610	0.108	-1252.403	123.654
region_southwest	-5.2385	313.322	-0.017	0.987	-620.072	609.596
sex_female	-364.4913	377.119	-0.967	0.334	-1104.514	375.531
sex_male	-118.6965	383.817	-0.309	0.757	-871.864	634.471
smoker_no	-1.207e+04	386.202	-31.255	0.000	-1.28e+04	-1.13e+04
smoker_yes	1.159e+04	407.239	28.454	0.000	1.08e+04	1.24e+04

Omnibus:	51.851	Durbin-Watson:	1.981
Prob(Omnibus):	0.000	Jarque-Bera (JB):	66.740
Skew:	0.481	Prob(JB):	3.22e-15
Kurtosis:	3.801	Cond. No.	5.90e+17

(b)

Fig. 3.14 Impact of removing outliers identified from cook's distance. (**a**) Model before. (**b**) Model after

often undefined. This section will discuss some techniques employed to visualize, validate, and diagnose clustering models.

One of the difficult choices in many clustering algorithms such as K-means, X-means, Expectation-Maximization, etc. is selecting number of clusters, usually symbolized by k. The choice depends on many factors such as the size of the data, dimensionality, end user's desire and prior knowledge. The optimal choice of k is a trade-off between maximum compression of the data and maximum separation between the unseen classes or the categories.

3.5.1 Elbow Method

For the elbow method, a clustering technique is run on the dataset for a range of values for k (say from 1–10). Then for each value of k, it computes an average distortion score for all the clusters. There are many ways to compute the distortion score; a common technique calculates the sum of square distances from each point to its assigned center. The plot of k and the average distortion score in a plot resembles the arm, then the k around the elbow, the point of inflection, is chosen as an optimum k. The elbow or the knee point is detected through an algorithm that finds the point of maximum curvature in the plot.

The Calinski-Harabasz score, also known as the Variance Ratio score, is the ratio of the sum of between-clusters dispersion and intercluster dispersion. The higher the Calinski-Harabasz score, the better is the clustering performance.

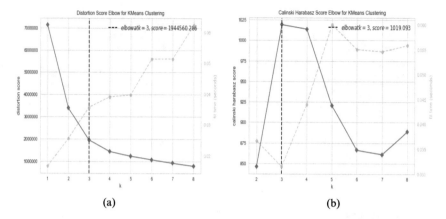

Fig. 3.15 Elbow Method for visualizing the optimum k for k-means clustering on the diabetes classification data. (**a**) Distortion scores. (**b**) Calinski-Harabasz score

Figure 3.15a and b show elbow detection using distortion and the Calinski-Harabasz method to find optimum k in the k-means for the diabetes classification data.

Observations:
Though the diabetes dataset has two labeled classes, both the distortion score and the Calinski-Harabasz score indicate that $k = 3$ is the optimum cluster size.

3.5.2   Silhouette Coefficient Visualizer

The Silhouette Coefficient is an estimate of the density of the clusters. It is computed for each instance based on two different scores as

- The mean distance between that instance and all other instances in the same cluster: a
- The mean distance between that instance and all other instances in the next nearest cluster: b

$$s = \frac{b - a}{max(b - a)} \qquad (3.2)$$

The Silhouette visualizer displays the silhouette coefficient for each instance on a per-cluster basis, visualizing the clusters and their density. Different plots for each value of k are shown in Fig. 3.16a, b, c, and d.

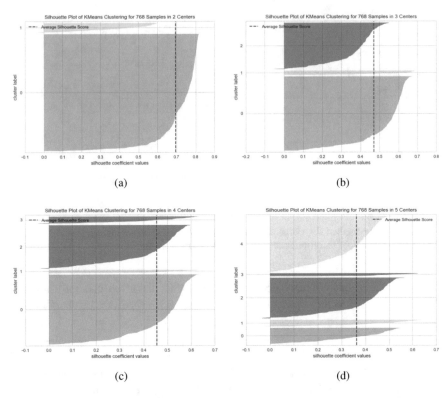

Fig. 3.16 Silhouette coefficients for k ranging from 2 to 5. (**a**) Silhouette coefficients for $k = 2$. (**b**) Silhouette coefficients for $k = 3$. (**c**) Silhouette coefficients for $k = 4$. (**d**) Silhouette coefficients for $k = 5$

> The Silhouette Coefficient has a best value of 1 and worst value of -1. Values near 0 indicate overlapping clusters. Negative values generally indicate that instances have the wrong cluster assignment.

Observations:
Based on the average Silhouette coefficient scores (indicated by red dotted line) for the diabetes dataset, the best k is 2 where the average score is high and there are no negative scores. The split between the two classes also seems to be in the same proportion as the original labeled class distribution.

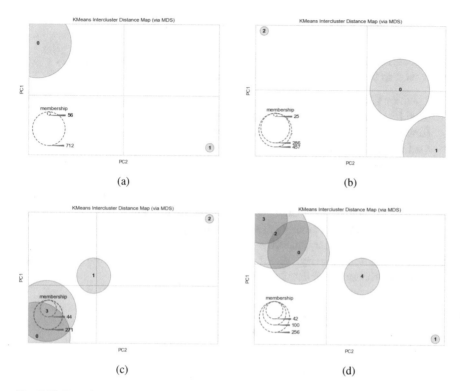

Fig. 3.17 Intercluster distance maps for different values of k. (**a**) $k = 2$. (**b**) $k = 3$. (**c**) $k = 4$. (**d**) $k = 5$

3.5.3   Intercluster Distance Maps

Intercluster distance maps visualize an embedding space in the lower dimensions of the cluster centers. Various projection techniques such as multidimensional scaling (mds), stochastic neighbor embedding (t-sne), etc. can be used for mapping from high dimensions to two dimensions. The clusters' memberships and sizes can be determined by a scoring method such as the number of instances belonging to each cluster and gives the clusters' relative importance (Fig. 3.17).

Observations:
Intercluster distance maps for various k using mds shows that for $k = 3$, based on the size, distribution and no overlaps indicate an ideal cluster size for the data.

3.6   Interpretable Machine Learning Properties

This section will detail some of the properties on which most algorithms can be compared from an interpretability standpoint.

1. **Local or Global**: Does the model provide interpretability at a single instance or local level or across the entire data space?
2. **Linearity**: Is the model capable of capturing non-linear relationships between the features?
3. **Monotonicity** Does the relationship between the feature and the target go in the same direction over the entire feature domain?
4. **Feature Interactions**: Some models capture interactions between the features while some assume independence. If captured in the right way, features interactions can increase the quality but simultaneously increase the complexity as well, thus reducing the interpretability.
5. **Best-suited Complexity**: Based on the hypothesis space of the model, what kinds of problem complexity is the algorithm best suited for?

3.7   Traditional Interpretable Algorithms

3.7.1   Tools and Libraries

Well-known open-source python packages like **statsmodels** and **sklearn** along with different data and plotting libraries were used for linear regression, logistic regression, Gaussian Naive Bayes, and Decision Tree. **pgmpy** is used for modeling Bayesian Network and **Orange** for Rule Induction.

3.7.2   Linear Regression

Linear regression is one of the oldest techniques that predicts the target using weights on the input features learned from the training data [KK62b]. The interpretation of the model becomes straightforward as the target is a linear combination of weights on the features. Thus linear regression model can be described as a linear combination of input $\mathbf{x}$ and a weight parameter $\mathbf{w}$ (that is learned during training process). In a d-dimensional input ($\mathbf{x} = [x_1, x_2, \ldots, x_d]$), we introduce another dimension called the bias term, x_0, with value 1. Thus the input can be seen as $\mathbf{x} \in \{1\} \times \mathbb{R}^d$, and the weights to be learned are $\mathbf{w} \in \mathbb{R}^{d+1}$.

In matrix notation, the input can be represented as a data matrix $\mathbf{X} \in \mathbb{R}^{N \times (d+1)}$, whose rows are examples from the data (e.g., $\mathbf{x}_n$), and the output is represented as a column vector $\mathbf{y} \in \mathbb{R}^N$. The process of learning via linear regression can be analytically represented as minimizing the squared error between the hypothesis function $h(\mathbf{x}_n)$ and the target real values y_n, as

$$E_{train}(h(\mathbf{x}, \mathbf{w})) = \frac{1}{N} \sum_{i=0}^{d} \left(\mathbf{w}^{\mathrm{T}}\mathbf{x}_n - y_n\right)^2 \qquad (3.3)$$

Since the data $\mathbf{x}$ is given, we will write the equation in terms of weights $\mathbf{w}$

$$E_{train}(\mathbf{w}) = \frac{1}{N} \|(\mathbf{Xw} - \mathbf{y})^2\| \qquad (3.4)$$

where $\|(\mathbf{Xw} - \mathbf{y})^2\|$ is the Euclidean norm of a vector.

This is an optimization problem that requires finding the weights $\mathbf{w}_{opt}$ that minimize the training error E_{train}.

$$\mathbf{w}_{opt} = \arg\min_{\mathbf{w} \in \mathbb{R}^{d+1}} E_{train}(\mathbf{w}) \qquad (3.5)$$

The solution for the weights is given by

$$\mathbf{w}_{opt} = \left(\mathbf{X}^{\mathrm{T}}\mathbf{X}\right)^{-1}\mathbf{X}^{\mathrm{T}}\mathbf{y} \qquad (3.6)$$

Linear regression makes the following assumptions that are important for model validation and interpretability

- Linearity: Linear regression assumes a linear relationship between the features and the label. In many real-world datasets this assumption may not hold true.
- Homoscedasticity: Linear regression assumes the error in the prediction will have a constant variance. This can be easily verified by plotting the results and looking at the scatter of the predictions from the linear hyperplane.
- Multicollinearity: If there is correlation between the features, the estimation of weights using linear regression is not accurate as the impact of the feature and its independence from others is lost.

Interpreting linear regression model can be summarized as below

- Increasing the continuous feature by one unit changes the estimated outcome by its weight.
- The categorical features should be transformed into multiple features. Each is encoded as a binary, 0 being the reference default and 1 is the presence of the feature. The interpretation for binary or categorical in such case

(continued)

is—changing the modified feature from the reference default to the other, changes the estimated outcome by the feature's weight.

- Intercept or the constant is the output when all the continuous features are at value 0 and the categories are in the reference default (e.g., 0). Understanding intercept value becomes meaningful for interpretation when the data is scaled with mean value 0 as it represents the default weight for an instance with mean values.
- Various regression methods such as Ordinary Least Squares (OLS) give not only the weights or the coefficients per feature but also standard error(std err), t-test(t), p-value(p) and the confidence intervals. The lower the standard error, the better is the accuracy of that coefficient and p-values less than a threshold alpha level indicate a statistically significant impact of that feature on the outcome.
- The R-squared value (also known as the coefficient of determination) provides a measure of how well the regression model explains the output value it is modeling. The closer the value is to 1.0, the better the model correctly describes the data.

Figure 3.18 gives the results of fitting a linear regression model on the claims insurance dataset.

There are various visualization techniques available for diagnosing or whiteboxing the regression. Figure 3.19 shows some of the known ways to analyze a feature *age* regressing with the output *charges*. Plot (a) which is the "Y and Fitted vs. X" graph plots the dependent variable against the predicted values with a confidence interval. Plot (b) shows the residuals of the model versus the chosen feature *age*. Each point in the plot is an observed value; the line represents the mean of those observed values. Plot (c) is the partial regression plot showing the relationship between the *charges* and the feature *age* conditional on the other independent features. The Component-Component plus Residual (CCPR) plot is an extension to the partial regression plot, a way to view the impact of one feature on the label by taking into account the effects of the other features. Thus it is $Res + w_i x_i$ versus x_i where Res is the residual of the whole model.

Explainable properties of linear regression are shown in Table 3.1.

3.7.2.1 Regularization

Regularization is a common technique employed in many weight-based learning methods to overcome the overfitting problem. There are many regularization techniques, of which we will highlight three of the most effective ones [HTF09b, HK00a].

OLS Regression Results

Dep. Variable:	charges		R-squared:	0.737
Model:	OLS		Adj. R-squared:	0.735
Method:	Least Squares		F-statistic:	371.7
Date:	Wed, 14 Apr 2021		Prob (F-statistic):	1.85e-301
Time:	21:23:04		Log-Likelihood:	-10851.
No. Observations:	1070		AIC:	2.172e+04
Df Residuals:	1061		BIC:	2.177e+04
Df Model:	8			
Covariance Type:	nonrobust			

	coef	std err	t	P>|t|	[0.025	0.975]
age	253.7005	13.530	18.751	0.000	227.152	280.249
bmi	335.9628	32.228	10.424	0.000	272.724	399.201
children	436.9101	156.584	2.790	0.005	129.661	744.159
region_northeast	380.4127	384.691	0.989	0.323	-374.430	1135.255
region_northwest	120.2800	376.809	0.319	0.750	-619.096	859.656
region_southeast	-532.8661	436.403	-1.221	0.222	-1389.177	323.445
region_southwest	-381.5360	391.071	-0.976	0.329	-1148.897	385.825
sex_female	-199.1229	471.328	-0.422	0.673	-1123.964	725.718
sex_male	-214.5866	477.395	-0.449	0.653	-1151.332	722.159
smoker_no	-1.201e+04	478.235	-25.112	0.000	-1.29e+04	-1.11e+04
smoker_yes	1.16e+04	510.457	22.716	0.000	1.06e+04	1.26e+04

Omnibus:	256.825	Durbin-Watson:	1.994
Prob(Omnibus):	0.000	Jarque-Bera (JB):	620.044
Skew:	1.279	Prob(JB):	2.29e-135
Kurtosis:	5.715	Cond. No.	4.39e+17

Fig. 3.18 Output of linear regression model on insurance dataset

Ridge regression or weight decay or L_2 norm is a regularization technique where less relevant features get weights close to 0 [HK00b]. The modified solution for regression can be written as

$$\mathbf{w}_{opt} = \underset{\mathbf{w} \in \mathbb{R}^{d+1}}{\arg\min} \left(E_{train}(\mathbf{w}) + \lambda \mathbf{w}^{\mathsf{T}} \mathbf{w} \right) \tag{3.7}$$

$$\mathbf{w}_{opt} = \left(\mathbf{X}^{\mathsf{T}} \mathbf{X} + \lambda \mathbf{I} \right)^{-1} \mathbf{X}^{\mathsf{T}} \mathbf{y} \tag{3.8}$$

where the regularization parameter λ is a hyperparameter and is generally a small value close to 0.

Lasso regression or L_1 norm is another popular regularization used in weight-based algorithms [HTF09b]. The modified equation for L_1 norm is

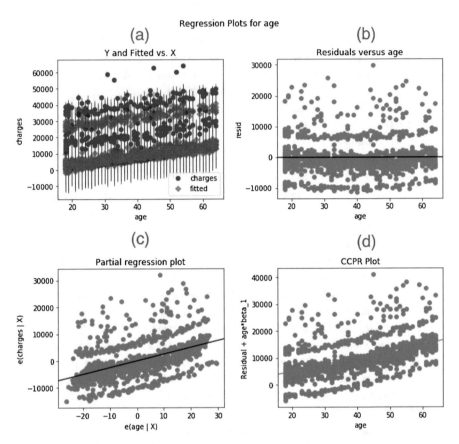

Fig. 3.19 Four different plots for feature age. (**a**) Regression plot showing fitted versus actual charges. (**b**) Residuals w.r.t age. (**c**) Partial regression plot and (**d**) CCPR plot

Table 3.1 Explainable
properties of Linear
regression

Properties	Values
Local or global	Global
Linear or non-linear	Linear
Monotonic or non-monotonic	Monotonic
Feature interactions captured	No
Model complexity	Low

$$\mathbf{w}_{opt} = \arg\min_{\mathbf{w} \in \mathbb{R}^{d+1}} \left(E_{train}(\mathbf{w}) + \lambda|\mathbf{w}| \right) \tag{3.9}$$

The absolute function in the above equation does not yield a closed-form solution and is represented as a constrained optimization problem as given below:

$$\arg\min_{\mathbf{w} \in \mathbb{R}^{d+1}} \left(\mathbf{X}^{T}\mathbf{X}\mathbf{w} - \mathbf{X}^{T}\mathbf{y} \right) \; s.t. \; \mathbf{w} < \eta \tag{3.10}$$

Table 3.2 Coefficients of features with basic OLS Regression, Lasso, Ridge and Elastic Net regularization

Features	LR	LR with Lasso	LR with Ridge	LR with ElasticNet
Age	253.70	253.70	252.17	256.53
Bmi	335.96	335.95	330.3	303.743
Children	436.91	436.86	439.24	411.71
Region_northeast	380.41	499.03	366.67	−6145.23
Region_northwest	120.28	238.98	98.59	−6410.03
Region_southeast	−532.86	−413.85	−464.24	−6917.32
Region_southwest	−381.53	−262.43	−395.66	−6862.1
Sex_female	−199.12	7.98	−203.56	−1.02
Sex_male	−214.58	−7.384	−191.08	−2.64
Smoker_no	−12, 010.0	−12, 335.45	−11, 645.20	−4829.02
Smoker_yes	11, 600.0	11, 270.25	11, 250.55	18, 764.06

where the hyperparameter η is inversely related to the regularization parameter λ.

Elastic Net combines both Lasso and Ridge regression [ZH03]. The modified equation is given by

$$\mathbf{w}_{opt} = \arg\min_{\mathbf{w} \in \mathbb{R}^{d+1}} \left(E_{train}(\mathbf{w}) + \lambda_1 |\mathbf{w}| + \lambda_2 \mathbf{w}^T \mathbf{w} \right) \tag{3.11}$$

Both L_2 and L_1 regularization can be seen as an implicit feature selection where the weights generally get reduced based on the relevance to the outcome but L_1 results in more feature weights being set to zero and thus a more sparse representation.

Table 3.2 shows how the feature weights change with different regularization techniques.

Observations:

- Figure 3.18 shows that features *age*, *bmi*, and *smoker_yes*, *smoker_no* all have *p*-values less than 0.005, indicating that they are statistically significant and thus their importance in predicting the insurance charges.
- Figure 3.18 also shows that the features *sex_male*, *sex_female* and various *region* have high *p*-values and can be considered not as significant and may be dropped for building models in an iterative way.

- Figure 3.18 highlights that the adjusted R-squared value is 0.735, and hence we can interpret it as: the model explains nearly 73.5% of the variation and can be considered a good fit.
- Figure 3.19 shows a linear relationship between *age* and *charges* with a positive trend, i.e., as the *age* increases the *charges* increase.
- Table 3.2 shows how every feature weight gets reduced with the introduction of regularization.

3.7.3 Logistic Regression

Linear regression is not practical on classification problems where the need is for the probability of the data belonging to a particular class rather than the linear interpolation between points. Logistic regression is a transformation θ applied on the linear combination $\mathbf{x}^T\mathbf{w}$ employed in the Linear Regression allowing a classifier to return a probability score [WD67].

$$h(\mathbf{x}) = \theta\left(\mathbf{w}^T\mathbf{x}\right) \tag{3.12}$$

A *logistic function* (also known as a *sigmoid* or *softmax* function) $\theta(\mathbf{w}^T\mathbf{x})$, shown below, is generally used for the transformation.

$$h(\mathbf{x}) = \frac{\exp \mathbf{w}^T\mathbf{x}}{1 + \exp \mathbf{w}^T\mathbf{x}} \tag{3.13}$$

For a binary classification, where $y \in \{-1, +1\}$, the hypothesis can be seen as a likelihood of predicting $y = +1$, i.e., $P(y = +1|\mathbf{x})$. Thus, the equation can be rewritten as an odds ratio, and weights are learned to maximize the *conditional likelihood* given the inputs.

$$\frac{P(y = +1|\mathbf{x})}{P(y = -1|\mathbf{x})} = \exp\left(\mathbf{w}^T\mathbf{x}\right) \tag{3.14}$$

Interpretation of a logistic regression model can be summarized as below

- Increasing the continuous feature by one unit changes estimated odds by a factor of $\exp(w_i x_i)$.

(continued)

- Similar to linear regression, the categorical features should be transformed into multiple features, with each encoded as a binary (0 being the reference default and 1 is the presence of that category) before modeling as a preprocessing step. Thus the interpretation is—when the categorical feature is changed from the reference category to the other category, the estimated odds change by a factor of $\exp(w_i x_i)$.
- The Intercept or the constant is the output when all the continuous features are at value 0 and the categories are at the reference default (e.g., 0). Thus, when all the continuous features have value 0, and the categorical features are in the default category, the intercept value gives the estimated odds.

Observations:

- Figure 3.20 shows weights or the coefficients for each feature. The value of 0.0311 for *Glucose* in the **coef** column means that for each unit increase in the value of *Glucose*, the log-odds of being classified as diabetic increases by a value of 0.0311. Also, higher glucose concentrations are positively associated with the diagnosis of diabetes.
- All the features except *BloodPressure* are positively associated with diagnosis of diabetes; as they increase, the log-odds of being classified as diabetic increases by the value in the **coef** column.
- The **P> |z|** column with alpha level of 0.05 shows features that are statistically significant in the classification. Features *Glucose, BMI, Pregnancies,* and *Insulin* can be considered statistically significant.

Explainable properties of logistic regression are shown in Table 3.3.

3.7.4 Generalized Linear Models

In Linear regression the continuous output is modeled as

$$y = w_0 + w_1 x_1 + \cdots + w_d x_d \tag{3.15}$$

with the assumption that the output y is normally distributed ($y \sim \mathcal{N}$) and the equation gives the expectation of the mean $\mathbb{E}(y)$ and with error/noise ϵ in $\mathcal{N}(0, \sigma^2)$.

```
Optimization terminated successfully.
         Current function value: 0.465382
         Iterations 6
```

Logit Regression Results

Dep. Variable:	Outcome	No. Observations:	614
Model:	Logit	Df Residuals:	605
Method:	MLE	Df Model:	8
Date:	Wed, 14 Apr 2021	Pseudo R-squ.:	0.2877
Time:	22:09:19	Log-Likelihood:	-285.74
converged:	True	LL-Null:	-401.18
Covariance Type:	nonrobust	LLR p-value:	1.946e-45

	coef	std err	z	P>|z|	[0.025	0.975]
const	-8.7245	0.891	-9.789	0.000	-10.471	-6.978
Pregnancies	0.0891	0.034	2.597	0.009	0.022	0.156
Glucose	0.0311	0.004	7.173	0.000	0.023	0.040
BloodPressure	-0.0098	0.010	-1.003	0.316	-0.029	0.009
SkinThickness	0.0284	0.015	1.875	0.061	-0.001	0.058
Insulin	0.0044	0.002	2.829	0.005	0.001	0.007
BMI	0.0646	0.020	3.276	0.001	0.026	0.103
DiabetesPedigreeFunction	0.7444	0.334	2.231	0.026	0.090	1.398
Age	0.0177	0.010	1.694	0.090	-0.003	0.038

Fig. 3.20 Logistic regression on the diabetes dataset

Table 3.3 Explainable properties of Logistic Regression

Properties	Values
Local or global	Global
Linear or non-linear	Linear
Monotonic or non-monotonic	Monotonic
Feature interactions captured	No
Model complexity	Medium

Generalized Linear Models (GLMs) have three basic components and relax the constraints or assumptions and generalize as the name suggests [MN89]. The three components are

1. The **distribution component**, which had an assumption of being normally distributed in the linear regression case, can be relaxed to be from any exponential family. Thus it can model skewed distributions.
2. The **linear predictor** is similar to linear regression and is linear in the weights trying to model the covariates.
3. The **link function** is the connection between the linear predictor and the mean of the distribution of the output or the label. In linear regression model the mean was equal to the linear predictor. In GLMs there can be a variety of link functions,

e.g., log of the means as the link function in the Poisson distribution or logit of the means for binomial logistic regression.

$$g(\mathbb{E}(y|\mathbf{x})) = \mathbf{w}_0 + \mathbf{w}_1\mathbf{x}_1 + \cdots + \mathbf{w}_d\mathbf{x}_d \qquad (3.16)$$

where g is the link function. Thus, GLM with Poisson distribution and log link function

$$\ln(\mathbb{E}(y|\mathbf{x})) = \mathbf{w}_0 + \mathbf{w}_1\mathbf{x}_1 + \cdots + \mathbf{w}_d\mathbf{x}_d \qquad (3.17)$$

Interpreting GLM can be summarized below:

- The distribution, along with the link function, suggests how to interpret the estimated feature weights. For example, in GLM with Poisson distribution and log as the link function, the output estimation is

$$\ln(\mathbb{E}(y|\mathbf{x})) = \mathbf{w}_0 + \mathbf{w}_1\mathbf{x}_1 + \cdots + \mathbf{w}_d\mathbf{x}_d \qquad (3.18)$$

and can be rewritten as

$$\mathbb{E}(y|\mathbf{x}) = \exp(\mathbf{w}_0 + \mathbf{w}_1\mathbf{x}_1 + \cdots + \mathbf{w}_d\mathbf{x}_d) \qquad (3.19)$$

So each feature contributes to the outcome ($\mathbb{E}(y)$) an exponential factor defined by the weight or the coefficient ($\exp(\mathbf{w}_i)$) multiplied by the exponential value of the feature ($\exp(\mathbf{x}_i)$).
- The positive or negative sign shows the increase or decrease in the exponential factor given the rest.
- The z and the **P**> |**z**| values give the test statistic and p-value, respectively, for the null hypothesis that a feature's regression coefficient is zero given that the rest of the features are in the model.

Observations:

- Figure 3.21 shows weights or the coefficients for each feature with Poisson Regression in GLM Model. The *age* coefficient of 0.02 is the Poisson regression estimate for a one unit increase in age, given the other features are held constant in the model. The interpretation is—if *age* were to increase by one unit, the difference in the log of expected value would be expected to increase by 0.02 unit, while holding the other features in the model constant.

Generalized Linear Model Regression Results

Dep. Variable:	charges	No. Observations:	1070
Model:	GLM	Df Residuals:	1061
Model Family:	Poisson	Df Model:	8
Link Function:	log	Scale:	1.0000
Method:	IRLS	Log-Likelihood:	-1.3395e+06
Date:	Fri, 11 Jun 2021	Deviance:	2.6673e+06
Time:	19:58:18	Pearson chi2:	3.37e+06
No. Iterations:	5		
Covariance Type:	nonrobust		

	coef	std err	z	P>|z|	[0.025	0.975]
age	0.0201	1.95e-05	1033.017	0.000	0.020	0.020
bmi	0.0261	4.43e-05	590.165	0.000	0.026	0.026
children	0.0400	0.000	181.126	0.000	0.040	0.040
region_northeast	1.6552	0.001	3034.622	0.000	1.654	1.656
region_northwest	1.6151	0.001	2934.801	0.000	1.614	1.616
region_southeast	1.5575	0.001	2537.661	0.000	1.556	1.559
region_southwest	1.5943	0.001	2754.425	0.000	1.593	1.595
sex_female	3.2133	0.001	4657.996	0.000	3.212	3.215
sex_male	3.2088	0.001	4563.059	0.000	3.207	3.210
smoker_no	2.5294	0.001	3536.012	0.000	2.528	2.531
smoker_yes	3.8927	0.001	5739.569	0.000	3.891	3.894

Fig. 3.21 Generalized linear model on the insurance dataset

- The positive coefficients for *age, bmi, children, region_* smoker_* and sex_* indicate the increase in the expected value of the charges with increase. The feature *smoker_yes* has the highest coefficient indicating the relevance of that feature in the regression model.
- All the features have 0.0 in the **P> |z|** column, thus all of them are statistically significant.

Explainable properties of GLM are shown in Table 3.4.

Table 3.4 Explainable properties of GLM

Properties	Values
Local or global	Global
Linear or non-linear	Linear
Monotonic or non-monotonic	Monotonic
Feature interactions captured	No
Model complexity	Low to medium

3.7.5 Generalized Additive Models

The assumption in all linear models is that the increase or decrease defined by the coefficient for that feature will be the same irrespective of the values. This assumption may not be true for many real-world applications where at the feature level, one may need a non-linear interaction. Generalized Additive Models (GAMs) are one of the ways to model the non-linear relationships by modifying GLMs [HT90a]. It is given by

$$g(\mathbb{E}(y|\mathbf{x})) = \mathbf{w}_0 + f_1(\mathbf{x}_1) + \cdots + f_d(\mathbf{x}_d) \qquad (3.20)$$

The equation generalizes the GLM equation where the generic function $f_i(\mathbf{x}_i)$ replaces the linear term $\mathbf{w}_i\mathbf{x}_i$. It gives the flexibility for non-linear interaction between the feature $\mathbf{x}_i$ and the output but still uses summation to capture overall feature impact. One easy way is to model the interactions as higher order polynomials at the feature level to capture non-linear relationship. **Splines** are piecewise polynomial curves, joining two or more polynomial curves, and can be generally used as the non-linear functions. A **smoothing spline** adds a constraint to the minimization problem such that the function $f(x_i)$ is twice differentiable and has a smoothing parameter λ that is like a penalty or regularization and the general equation for minimization is given as

$$MSE = \frac{1}{N} \sum_{i=0}^{n} (y_i - f(x_i))^2 + \lambda \int f''(x)^2 dx \qquad (3.21)$$

The output from a GAM is less interpretable as it does not have coefficients like others but λ values for different feature fits as shown in Fig. 3.22. Normally, partial dependence plot (which is graphical), where the output is plotted against the fitted function for a feature as shown in Fig. 3.23, is used to understand individual feature mappings to the non-linear or linear functions.

```
LinearGAM
==================================================  ==================================================================
Distribution:                   NormalDist Effective DoF:                                              29.0
Link Function:                IdentityLink Log Likelihood:                                      -19629.2742
Number of Samples:                    1070 AIC:                                                 39318.5484
                                           AICc:                                                39320.3386
                                           GCV:                                              38882924.9935
                                           Scale:                                            36990717.0651
                                           Pseudo R-Squared:                                         0.749
==================================================  ==================================================================
Feature Function          Lambda              Rank         EDoF         P > x         Sig. Code
================          ===========         =========    ==========   ===========   ============
s(0)                      [0]                 10           10.0         1.11e-16      ***
s(1)                      [0]                 10           9.0          1.11e-16      ***
s(2)                      [0]                 10           5.0          1.11e-16      ***
s(3)                      [0]                 10           1.0          1.11e-16      ***
s(4)                      [0]                 10           1.0          1.11e-16      ***
s(5)                      [0]                 10           1.0          1.11e-16      ***
s(6)                      [0]                 10           0.0          1.11e-16      ***
s(7)                      [0]                 10           1.0          1.11e-16      ***
s(8)                      [0]                 10           0.0          1.11e-16      ***
s(9)                      [0]                 10           0.9          1.11e-16      ***
s(10)                     [0]                 10           0.0          1.11e-16      ***
intercept                                     1            0.0          3.62e-04      ***
==================================================  ==================================================================
Significance codes:  0 '***' 0.001 '**' 0.01 '*' 0.05 '.' 0.1 ' ' 1
```

Fig. 3.22 Linear generalized additive models on the insurance dataset

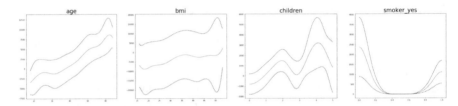

Fig. 3.23 Linear generalized additive partial dependence plots for features age, bmi, children and smoker with mean and 95% confidence interval bounds

---

Observations:

- The preprocessed dataset is reduced to only four features – $age, bmi, children, smoker\_yes$.
- Linear GAM with 10 splines is chosen after doing a grid search for various linear, splines, and factor terms.
- Figure 3.22 shows fitting various functions for each feature and their statistical significance in the **Sig. Code** column. Every feature function is statistically significant.
- The partial dependence plot as shown in Fig. 3.23 show how splines actually capture the non-linear relationships in a smooth way especially for *children*. It also shows how the *age*, *children*, and *bmi* have positive correlation to the insurance charges and thus provides the needed explainability.

Explainable properties of GAM are shown in Table 3.5.

Table 3.5 Explainable
properties of GAM

Properties	Values
Local or global	Global
Linear or non-linear	Non-linear
Monotonic or non-monotonic	Monotonic
Feature interactions captured	No
Model complexity	Low to medium

3.7.6 Naive Bayes

Naive Bayes is one of the simplest algorithms based on the Bayes theorem [RN09]. There are many advantages to Naive Bayes, such as simplicity, explainability, speed, and ability to learn from few examples. The hypothesis in general Bayes equation for a binary classification $y_i \in (0, 1)$ is given by

$$h_{Bayes}(\mathbf{x}) = \arg\max_{y \in (0,1)} P(X = \mathbf{x}|Y = y)P(Y = y) \tag{3.22}$$

In *Naive Bayes*, there is an assumption of independence between the features. So, for d dimensions, the equation simplifies as

$$h_{Bayes}(\mathbf{x}) = \arg\max_{y \in (0,1)} P(Y = y) \prod_{j=1}^{d} P(X_j = x_i|Y = y) \tag{3.23}$$

As a result, training and estimating parameters of Naive Bayes just measures two quantities, the priors for the class $P(Y = y)$ and the conditional for each feature $P(X_j = x_j|Y = y)$ given the class or the label.

A dataset that has continuous features can be discretized using many known techniques [Gar+12]. Also, many implementations also assume a Gaussian distribution and the probability distribution is given by

$$P(X = \mathbf{x}|Y = y) = \frac{1}{\sqrt{2\pi \sigma_I k^2}} e^{\frac{(x - \mu_i k)^2}{\sigma_i k^2}} \tag{3.24}$$

Figure 3.24 shows output of Gaussian Naive Bayes for the diabetes dataset, where the mean and standard deviations for each class are estimated from the training data for each feature and class.

The independence assumption in Naive Bayes contributes to its simplicity and interpretability.

	Feature	Diabetic Variances	Diabetic Means	Non-Diabetic Means	Non-Diabetic Variances
0	Pregnancies	14.922021	4.764706	3.374046	9.409716
1	Glucose	854.767314	141.628959	110.493639	601.766507
2	BloodPressure	152.052198	75.110860	71.231552	134.529089
3	SkinThickness	78.039647	32.357466	27.394402	71.994583
4	Insulin	8746.455363	186.805430	119.050891	6410.156452
5	BMI	44.746775	35.566063	30.940967	43.563088
6	DiabetesPedigreeFunction	0.141134	0.538986	0.425692	0.089203
7	Age	123.166815	37.420814	31.442748	143.091514

Fig. 3.24 Gaussian Naive Bayes on diabetes dataset

Interpreting Naive Bayes model can be summarized as below

- The means and standard deviations for each class per feature can be used to compute the probabilities for that class and feature. For example, if we want to see predictions for an instance with $\{Pregnancies = 2, Glucose = 120.2, BloodPressure = 75.38, SkinThickness = 25.18, Insulin = 121.75, DiabetesPedigreeFunction = 0.75, Age = 34\}$, we will compute the posterior for each feature per class. The probability for feature $Glucose$ for each class is computed by plugging in the values from Fig. 3.24 as

$$P(X = 120.02 | Y = 0) = \frac{1}{\sqrt{2\pi 601.76^2}} e^{\frac{(120.0 - 110.49)^2}{601.76^2}} \tag{3.25}$$

and

$$P(X = 120.02 | Y = 1) = \frac{1}{\sqrt{2\pi 854.76^2}} e^{\frac{(120.0 - 141.62)^2}{854.76^2}} \tag{3.26}$$

All the probabilities are then multiplied and the class is selected based on the maximum value.
- Thus the output of Naive Bayes and how each feature contributes to the prediction based on the probability values is easily interpretable.

Explainable properties of Naive Bayes are shown in Table 3.6.

Table 3.6 Explainable
properties of Naive Bayes

Properties	Values
Local or global	global
Linear or non-linear	Non-linear
Monotonic or non-monotonic	Monotonic
Feature interactions captured	No
Model complexity	Low to medium

3.7.7 Bayesian Networks

In traditional machine learning, encoding an expert's knowledge requires labor-intensive feature engineering. Understanding causality from how a feature influences the outcome and the ability to map model outputs to capture uncertainty, both require rigorous analysis and, most often, surrogate methods are used.

Bayesian networks are probabilistic graphical models (PGM) that use Bayesian inference to model an expert's knowledge and uncertainty from the data. Bayesian networks aim to model conditional dependence between the features and, therefore, can capture causality [Pea88, CY95, FGG97]. Bayesian networks satisfy the local Markov property, i.e., a node is conditionally independent of its non-descendants given its parents. Bayesian networks has

- a set of nodes (features observed or unobserved),
- a directed, acyclic graph (edges between nodes are "direct influences"I), and
- a conditional distribution for each node given its parents

Thus, the joint distribution for a Bayesian network is equal to the product of P(node|parents(node)) for all nodes, stated below:

$$P(X_1, \cdots, X_n) = \prod_{i=1}^{n} P(X_i | X_1, \cdots, X_{i-1}) - \prod_{i=1}^{n} P(X_i | Parents(X_i))$$

$$(3.27)$$

There are many algorithms to perform inferencing in Bayesian networks. Exact methods like variable elimination take advantage of the fact that each factor only involves a small number of features and work very efficiently for a small number of features. As the number of features increase it becomes computationally infeasible to perform inferencing through direct methods and approximate methods like Markov Chain Monte Carlo (MCMC) are used [KF09].

In classification problems, data is either discretized to calculate the conditional probabilities or parameterized distributions such as Gaussians are used for continuous features.

For the diabetes dataset, the Bayesian network is constructed with domain knowledge of how certain features influence others and the outcome [GBH12].

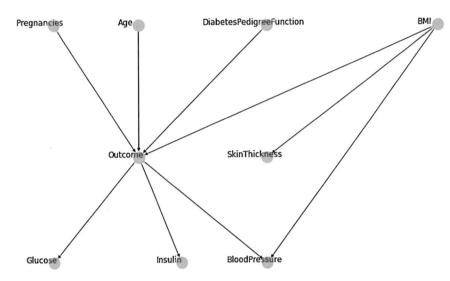

Fig. 3.25 Bayesian network for diabetes dataset

	precision	recall	f1-score	support
0	0.780	0.868	0.821	53
1	0.611	0.458	0.524	24
accuracy			0.740	77
macro avg	0.695	0.663	0.673	77
weighted avg	0.727	0.740	0.729	77

Fig. 3.26 Bayesian prediction results on the test set

Observations:

- Figure 3.25 indicates how features *Pregnancies*, *Age*, *DiabetesPedigree-Function* directly influence the outcome *diabetes*. The research modeled an unobserved variable "overweight" that influences *SkinThickness*, *BMI*, and outcome [GBH12]. But we changed the graph by mapping the feature *BMI* directly to the overweight variable and influencing *SkinThickness* and the outcome.
- Instead of assuming any parameterized distribution such as Gaussian, we use binning and discretize all the continuous features.
- Figure 3.26 shows the performance of the model on test data sampled from the data and has relatively good precision/recall as compared to other methods with the advantage of high interpretability.

Explainable properties of Bayesian Networks are shown in Table 3.7.

Table 3.7 Explainable properties of Bayesian Networks

Properties	Values
Local or global	Global and local
Linear or non-linear	Non-linear
Monotonic or non-monotonic	Non-monotonic
Feature interactions captured	Yes
Model complexity	Medium

3.7.8 Decision Trees

Decision Trees are the most popular interpretable algorithm for classification and regression. The general idea is to construct a binary tree with a decision point on the feature's value as a cut-off where the tree branches and splits the data. Based on how to choose features at the splitting, different ways to split the feature based on values, how deep to grow, how to reduce the tree's size, etc., there are many decision trees variants.

Classification and Regression Trees (CART) is one of the most popular decision tree algorithms which employs the Gini index metric to decide which feature to split the tree on [Bre+84]. Gini index is a measure of impurity and for k classes is measured as

$$Gini = 1 - \sum_j p_j^2 \qquad (3.28)$$

There are a number of other techniques such as entropy, classification error, etc. that have been employed successfully.

Interpreting Decision Tree model can be summarized as below

- As shown in Fig. 3.27, every decision tree node has the splitting feature and threshold (e.g., *Insulin* $\leq$ 121.0), splitting metric value (e.g., Gini value of 0.461 at the root), and population in each class ([393, 221]).
- As discussed, Gini score quantifies the purity of the node/leaf. A Gini score greater than zero implies that samples contained within that node belong to different classes. A Gini score of zero means that the node is pure, i.e., that node consists of representatives from only one class.
- Starting from the root node and traversing all the way to leaves, various human-interpretable rules can be derived. For example, *Insulin* $\leq$ $121.0 AND Glucose \leq 151.5 AND Pregnancies \leq 14.0$ is a predictor of *non-diabetes* with 302 samples and resulting in only 15 errors (diabetes).
- The decreasing Gini score at each node level shows why the node/leaf is getting purer, and the rules are generalized.

Explainable properties of CART are shown in Table 3.8.

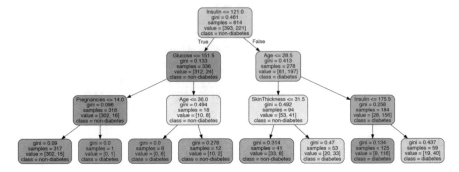

Fig. 3.27 Decision Tree which is constrained to be three-level deep for the diabetes classification data

Table 3.8 Explainable properties of CART

Properties	Values
Local or global	Global
Linear or non-linear	Non-linear
Monotonic or non-monotonic	Some monotonicity
Feature interactions captured	Yes
Model complexity	Medium to high

3.7.9 Rule Induction

Rule induction is another popular traditional white-box technique in machine learning. Instead of starting from decision trees and converting them into rules, rule induction induces rules in the form of "IF <conditions >then class." As compared to decision trees, which use the "divide-and-conquer" strategy, rule induction works through the "separate-and-conquer" approach [BGH89, CG91, Mic83a].

The general algorithm is to learn "one rule" at a time that "covers" positive instances in the dataset, remove those, and iteratively learn new rules until all positives are covered. The technique is also known as "sequential covering." Creating conditions for the "if" requires searching for feature-value combinations, and there are various search techniques such as exhaustive, greedy/heuristic-based such as beam search, genetic algorithms, etc. There are multiple metrics to evaluate while learning a rule, such as accuracy, weighted accuracy, precision, information gain, etc., thus resulting in many variants. Similarly, there are multiple ways to arrange the rules during the inference. One can order the rules in the same way that it learned, metrics-based (accuracy, etc.), or some strategy-based for unordered execution. Often, the rules, like decision trees, can overfit to the training data. Two general approaches to overcome overfitting are pre-pruning and post-pruning. In pre-pruning, the rules stop at a certain point before it classifies or covers the instances perfectly, thus introducing some errors. In post-pruning, the training data is further split into growing and pruning sets; rule learning happens on the growing set to overfit the data, and post-pruning prunes these rules and uses the pruning set as validation data.

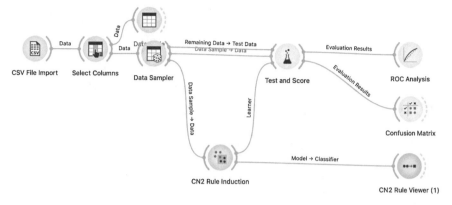

Fig. 3.28 Building CN2 rule induction using orange

The most popular rule induction algorithms used in many applications are CN2, M5Rules, and RIPPER [CN89, Coh95a, HHF99, Qui92]. We use the Orange library to model the diabetes dataset as shown in Fig. 3.28.

Interpretation of CN2 Rules:

- CN2 output has a sequence of ordered rules. Each rule has an "If condition" part that has at minimum a triplet of feature, operator, and value, e.g., *Glucose* $\geq$ 158.0 or combinations of these triplets with "AND" operator, e.g., *Glucose* $\geq$ 158.0 *AND SkinThickness* $\geq$ 44.0.
- The rule also has the "THEN clause" that implies a class that the rule captures (positive or negative in binary classification) and the distribution of positives and negatives the rule captures.
- The negatives captured in the positive class are the false positives, and the positives captured in the negative class are the false negatives.

Observations:

- We constrained the CN2 algorithm to have maximum rule length of 5, i.e., not more than 5 feature-operator-value are in conjunction. We also constrain that a rule should at least capture 8 examples in the dataset. These hyperparameters were manually searched. The maximum rule length and minimum examples act as a regularizer and prevent overfitting.

IF conditions		THEN class ⌄	Distribution
Glucose≥158.0 AND SkinThickness≥44.0	→	Outcome=1	[0, 8]
Age≤29.0 AND Glucose≥171.0	→	Outcome=1	[0, 13]
BMI≥45.4 AND BloodPressure≥80.0 AND SkinThickness≥36.0	→	Outcome=1	[0, 9]
Pregnancies≥7.0 AND Glucose≥155.0 AND Glucose≥181.0	→	Outcome=1	[0, 12]
BloodPressure≥30.0 AND Glucose≥131.0	→	Outcome=1	[0, 8]
Insulin≥205.0 AND Pregnancies≥8.0 AND Age≥36.0	→	Outcome=1	[0, 8]
DiabetesPedigreeFunction≥0.731 AND Pregnancies≥8.0 AND Age≥42.0	→	Outcome=1	[0, 8]
Pregnancies≥7.0 AND Age≤33.0 AND BloodPressure≥66.0	→	Outcome=1	[0, 9]
Glucose≥155.0 AND Age≤31.0	→	Outcome=1	[1, 12]
BloodPressure≥90.0 AND DiabetesPedigreeFunction≤0.674 AND Glucose≥137.0	→	Outcome=1	[0, 10]
DiabetesPedigreeFunction≥0.855 AND BMI≤40.0 AND BloodPressure≥66.0 AND Glucose≥109.0	→	Outcome=1	[0, 11]
BMI≥31.6 AND BMI≤33.1 AND Glucose≥144.0	→	Outcome=1	[0, 9]
Age≥43.0 AND BloodPressure≤80.0 AND Pregnancies≥6.0 AND Glucose≤154.0 AND DiabetesPedigreeFunction≥0.257	→	Outcome=1	[0, 8]
Insulin≥70.0 AND Insulin≤193.0 AND BMI≥34.2 AND BloodPressure≥60.0	→	Outcome=1	[0, 14]
DiabetesPedigreeFunction≥0.484 AND BloodPressure≥70.0 AND Glucose≥116.0	→	Outcome=1	[0, 9]
Glucose≥158.0	→	Outcome=1	[2, 9]
DiabetesPedigreeFunction≤0.28 AND Glucose≥123.0 AND DiabetesPedigreeFunction≥0.205 AND BloodPressure≥70.0	→	Outcome=1	[0, 8]
Glucose≥112.0 AND Glucose≤117.0 AND BloodPressure≥60.0	→	Outcome=1	[0, 8]
Age≥22.0 AND BloodPressure≥74.0 AND Age≥32.0 AND DiabetesPedigreeFunction≥0.259	→	Outcome=1	[0, 8]
SkinThickness≥36.0 AND DiabetesPedigreeFunction≥0.337 AND Age≥22.0	→	Outcome=1	[1, 8]
Age≥24.0 AND DiabetesPedigreeFunction≤0.243	→	Outcome=1	[1, 7]
Age≥24.0 AND Glucose≥95.0	→	Outcome=1	[1, 7]

Fig. 3.29 Rules covering positive class in the diabetes dataset

Table 3.9 Explainable properties of CN2 Rules

Properties	Values
Local or global	Global and local
Linear or non-linear	Non-linear
Monotonic or non-monotonic	Some monotonicity
Feature interactions captured	Yes
Model complexity	Medium to high

- Figure 3.29 shows only the rules for the positive class, i.e., outcome = 1.
- The CN2 Rule Induction algorithm generates 55 rules on the dataset, 22 for the positive class.
- Only 4 rules out of 22 generate false positives, showing a good recall on the training data.
- There are interesting domain-specific rules such as "*Age* $\leq$ 31 *AND Glucose* $\geq$ 155.0" and "*Age* $\leq$ 29 *AND Glucose* $\geq$ 171.0" which captures young population with high glucose.
- There are some interesting ranges of certain features and relationship with other features captured such as the rule "*Insulin* $\geq$ 70 *AND Insulin* $\leq$ 193.0 *AND BMI* $\geq$ 34.2 *AND BloodPressure* $\geq$ 60.0" with 14 true positives with no false positives.

Explainable properties of CN2 rules are shown in Table 3.9.

References

[Bel+80]   D.A. Belsley, et al., *Regression Diagnostics: Identifying Influential Data and Sources of Collinearity*. Wiley Series in Probability and Statistics - Applied Probability and Statistics Section Series (Wiley, Hoboken, 1980). ISBN: 9780471058564

[BGH89]   L.B. Booker, D.E. Goldberg, J.H. Holland, Classifier systems and genetic algorithms. Artif. Intell. **40**(1–3), 235–282 (1989)

[Bra97]   A.P. Bradley, The use of the area under the ROC curve in the evaluation of machine learning algorithms. Pattern Recog. **30**(7), 1145–1159 (1997)

[Bre+84]   L. Breiman et al., *Classification and Regression Trees* (Wadsworth and Brooks, Monterey, 1984)

[CT10]   G.C. Cawley, N.L.C. Talbot, On over-fitting in model selection and subsequent selection bias in performance evaluation. J. Mach. Learn. Res. **11**, 2079–2107 (2010)

[CG91]   C.-C. Chan, J.W. Grzymala-Busse, *On the attribute redundancy and the learning programs ID3, PRISM, and LEM2. Department of Computer Science, University of Kansas*. Technical Report TR-91- 14, December 1991 (1991)

[Che+05]   J. Chen et al., *The Use of Decision Threshold Adjustment in Classification for cancer Prediction* (National Center for Toxicological Research Food and Drug Administration, Jefferson, Arkansas, 2015). http://www.ams.sunysb.edu/~hahn/psfile/papthres.pdf

[CN89]   P. Clark, T. Niblett, The CN2 induction algorithm. Mach. Learn. **3**(4), 261–283 (1989)

[Coh95a]   W.W. Cohen, Fast effective rule induction, in *Machine Learning Proceedings 1995* (Elsevier, Amsterdam, 1995), pp. 115–123

[Coo11]   R.D. Cook, Cook's distance, in *International Encyclopedia of Statistical Science* (Springer, Berlin, 2011), pp. 301–302. ISBN: 978-3-642-04898-2

[CY95]   G.F. Cooper, C. Yoo, Causal discovery from a mixture of experimental and observational data, in *UAI '99: Proceedings of the Fifteenth Annual Conference on Uncertainty in Artificial Intelligence* (Morgan Kaufmann, Burlington, 1995), pp. 116–125

[DG06]   J. Davis, M. Goadrich, The relationship between precision- recall and ROC curves, in *ICML '06: Proceedings of the 23rd International Conference on Machine Learning* (Association for Computing Machinery, New York, 2006), pp. 233–240. ISBN: 1-59593-383-2

[FGG97]   N. Friedman, D. Geiger, M. Goldszmidt, Bayesian network classifiers. Mach. Learn. **29**(2–3), 131–163 (1997)

[Gar+12]   S. Garcia, et al., A survey of discretization techniques: Taxonomy and empirical analysis in supervised learning. IEEE Trans. Knowl. Data Eng. **25**(4), 734–750 (2012)

[GV18]   T. Gneiting, P. Vogel, *Receiver Operating Characteristic (ROC) Curves* (2018). arXiv: 1809.04808 [stat.ME]

[GBH12]   Y. Guo, G. Bai, Y. Hu, Using bayes network for prediction of type-2 diabetes, in *2012 International Conference for Internet Technology and Secured Transactions* (IEEE, Piscataway, 2012), pp. 471– 472

[HT90a]   T.J. Hastie, R.J. Tibshirani, *Generalized Additive Models*, vol. 43 (CRC Press, Boca Raton, 1990)

[HTF09b]   T. Hastie, R. Tibshirani, Generalized additive models: some applications. J. Amer. Statist. Assoc. **82**(398), 371–386 (1987)

[HTF09a]   T. Hastie, R. Tibshirani, J. Friedman, *The Elements of Statistical Learning*. Springer Series in Statistics, Chap. 15 (Springer, Berlin, 2009)

[HK00a]   A.E. Hoerl, R.W. Kennard, Ridge regression: biased estimation for nonorthogonal problems. Technometrics **42**(1), 80–86 (2000). ISSN: 0040-1706. http://doi.org/10.2307/1271436

[HK00b]   M. Hollander, D.A. Wolfe, *Nonparametric Statistical Methods* (Wiley, New York, 1973)

[HHF99]   G. Holmes, M. Hall, E. Frank, Generating rule sets from model trees, in *Twelfth Australian Joint Conference on Artificial Intelligence* (Springer, Berlin, 1999), pp. 1–12

[KK62b]   J.F. Kenney, E.S. Keeping, *Mathematics of Statistics*. (van Nostrand, Princeton, 1962), pp. 252–285

[KF09]   D. Koller, N. Friedman, *Probabilistic Graphical Models: Principles and Techniques* (MIT Press, Cambridge, 2009)

[MN89]   P. McCullagh, J.A. Nelder, *Generalized Linear Models*. Chapman and Hall/CRC Monographs on Statistics and Ap- plied Probability Series, 2nd edn. (Chapman & Hall, London, 1989). ISBN: 9780412317606. http://books.google.com/books?id=h9kFH2%5C_FfBkC

[Mic83a] R.S. Michalski, A theory and methodology of inductive learning, in *Machine Learning* (Elsevier, Amsterdam, 1983), pp. 83–134

[Pea88]   J. Pearl, *Probabilistic Reasoning in Intelligent Systems: Networks of Plausible Inference* (Morgan Kaufmann, Burlington, 1988)

[Per10]   C. Perlich, Learning curves in machine learning, in *Encyclopedia of Machine Learning* (Springer US, Berlin, 2010). ISBN: 978-0-387-30164-8

[Pro]   F. Provost, *Machine Learning from Imbalanced Data Sets 101* (Technical Report WS-00-05, AAAI, Menlo Park, CA, 2000), pp. 1–3

[Qui92]   R.J. Quinlan, Learning with continuous classes, in *5th Australian Joint Conference on Artificial Intelligence* (World Scientific, Singapore, 1992), pp. 343–348

[Ras20]   S. Raschka, *Model Evaluation, Model Selection, and Algorithm Selection in Machine Learning* (2020). arXiv: 1811.12808 [cs.LG]

[RN09]   S.J. Russell, P. Norvig, *Artificial Intelligence: A Modern Approach*, 3rd edn. (Pearson, London, 2009)

[WD67]   S.H. Walker, D.B. Duncan, Estimation of the probability of an event as a function of several independent variables. Biometrika **54**, 167–179 (1967)

[ZH03]   H. Zou, T. Hastie, Regularization and variable selection via the elastic net. J. Roy. Statist. Soc. Ser. B (Statist. Methodol.) **67**(2), 301–320 (2003)

Chapter 4
Model Interpretability: Advances in Interpretable Machine Learning

This chapter expands on intrinsic model interpretability discussed in the last chapter to include many modern techniques that are both interpretable and accurate on many real-world problems. The chapter starts with differentiating between interpretable and explainable models and why, in specific domains where high stakes decisions need to be made, interpretable models should be a natural choice. The chapter covers some state-of-the-art interpretable models that are ensemble-based, decision tree-based, rules-based, and scoring system based. We describe each algorithm in sufficient detail and then use the diabetes classification or insurance claims regression dataset to practically demonstrate the output of each, along with interpretations and observations.

4.1 Interpretable vs. Explainable Algorithms

In the paper *Stop Explaining Black-Box Machine Learning Models for High Stakes Decisions and Use Interpretable Models Instead*, Cynthia Rudin differentiates between the explainable and interpretable models [Rud19b]. The paper cites examples in real-world domains such as unfair credit loan rejection, discriminating bail and parole rejection, wrong medical diagnosis, etc. caused by black-box machine learning models. Many techniques have evolved that enhance the explainability of black-box machine learning models rather than making models more interpretable. The paper makes the case that building post-hoc techniques can have a lasting negative impact on the widespread use of machine learning models, especially in high stakes decisions and model troubleshooting. Next, we discuss some of the issues with explainable methods on black-box models, as discussed by Rudin.

1. The trade-off between accuracy and interpretability is a myth and the demonstration of various interpretable algorithms by comparing them with black-box models for accuracy in different domains confirms this [Rud19c]. By using

U. Kamath, J. Liu, *Explainable Artificial Intelligence: An Introduction to Interpretable Machine Learning*, https://doi.org/10.1007/978-3-030-83356-5_4

interpretable models instead of explainable models, you can understand the issues with the data and model more effectively which can then be corrected by following the iterative KDD or CRISP-DM methods [AS08].

2. Explanation methods as a post-hoc on black-box models are not 100% faithful to the original, and hence the approximations can cause trust issues. Using explanation techniques outside the original black-box results in two logical models that one relies on instead of just the original model. If the explanation methods are 90% accurate, that leads to at least 10% inaccurate cases and may prove costly, especially in a high stakes decision.

3. Explanation methods often do not provide enough detail to understand how the black-box models are predicting. For example, saliency maps are explainable techniques used in computer vision that indicate what part of the image is being omitted or seen by the classifier, but that does not provide any information about what the model is doing [Ade+18].

4. Sometimes, other factors outside the data on which the model is based need to be considered, for example, circumstances of the crime in criminal justice rather than just the crime-related data that models bails and paroles. Not having an interpretable or transparent model makes this a difficult task.

5. Explainable methods can give rise to a complex decision pathway to explain the black-box model, leading to more human errors.

On the other hand, constructing interpretable models requires solving complex statistical hurdles such as creating optimal global trees, solving hard constraint optimization problems, and implementing systemic software enhancements such as caching and data structures for improved performance. We have broadly classified the algorithms into the following areas, viz. (1) Ensemble-based, (2) Rules or decision list, (3) Decision tree-based, and (4) Scoring systems-based. The rest of the chapter expands on each of these areas.

4.2   Tools and Libraries

Table 4.1 provides details of all the libraries used for various models in the chapter.

4.3   Ensemble-Based

One of the ensemble-based techniques is to build models that combine the outputs given by a collection of trees, as opposed to a single one, such as Random Forests and Gradient Boosting Machines [Bre01, Fri00]. Boosting is a sequential ensemble process where you combine the classifiers by putting more weight on the observations misclassified in the previous step [FSA99]. Bayesian model averaging is another ensemble approach where the set of tree models have prior

Table 4.1 Models and implementations

Model/algorithm	Library
Boosted rulesets	https://github.com/csinva/imodels
Explainable boosting machine	https://github.com/interpretml/interpret
RuleFit	https://github.com/csinva/imodels
Skope-Rules	https://github.com/csinva/imodels
Iterative random dom forest	https://github.com/Yu-Group/iterative-Random-Forest
Optimal classification trees	https://github.com/pan5431333/pyoptree
Optimal decision trees (branch and bound)	https://github.com/aia-uclouvain/pydl8.5
Optimal sparse decision trees	https://github.com/xiyanghu/OSDT
Generalized and scalable optimal decision tree	https://github.com/Jimmy-Lin/GeneralizedOptimalSparseDecisionTrees
Bayesian ors of ands	https://github.com/wangtongada/BOA
Bayesian case model	https://users.cs.duke.edu/~cynthia/code.html
Certifiably optimal RulE ListS	https://github.com/corels/corels
Sparse linear integer models	https://github.com/csinva/imodels
Risk-calibrated supersparse linear integer model	https://github.com/ustunb/risk-slim

distributions, and some stochastic search techniques find the good tree models from them [Was+00, CGM+10]. Many of these ensemble-based techniques are uninterpretable and rely on explainable methods such as variable importance and partial dependence plots for explanation.

4.3.1 Boosted Rulesets

One of the common ways to build interpretable classifiers using ensemble techniques is to combine a weak classifier such as a decision stump and use the AdaBoost algorithm [FS97]. The weak learning algorithm A iteratively produces a collection of weak classifiers, and a linear combination of these results in a strong classifier. AdaBoost produces a discrete probability distribution over the instances. The harder to classify instances in every iteration, i.e., those previously misclassified, get higher weight using an exponential weighting scheme. The algorithm can be summarized below as:

Algorithm 1: AdaBoost

Data: training data $S = \{x_i, y_i\}_{i=1}^m$, where $y_i \in \{-1, 1\}$

Initialize $d_{1,i} = 1/m \; for \; i = 1, \cdots, m$

for $t = 1, \cdots, T$ **do**

   Train Weak learner, producing $h(t) : X \longrightarrow \{-1, 1\}$

   Calculate error $\epsilon_t = P_{i \sim d_t}[h_{(t)}(x_i) \neq y_i]$

   Calculate coefficients $\alpha_t = \frac{1}{2}\left(\frac{1-\epsilon_t}{\epsilon_t}\right)$

   Update weights $d_{t+1,i} = \frac{d_{t,i}}{Z_t} e^{-y_i \alpha_t h_{(t)} x_i}$

return $H = \text{sign}\left(\sum_{t=1}^{T} \alpha_t h_{(t)}\right)$

Interpretation of AdaBoost with decision stump: AdaBoost outputs various decision rules where each rule will be of type (*feature condition value*), the *condition* will be $\{\leq, <, =, >, \geq\}$ for continuous features and $\{=, \neq\}$ for categorical features. The *value* will be a continuous value or a specific category based on continuous and categorical features.

Observations:

- Figures 4.1 and 4.2 show most features *Insulin, Glucose, Age, SkinThickness, BMI, BloodPressure* are discriminating and have some meaningful thresholds in tree of size 1 and 2.
- Figures 4.3 and 4.4 showing the precision-recall curve at various thresholds indicate that the rules generated with simple size 1 are more generic and have better metrics.

Explainable properties of Boosted Rulesets are shown in Table 4.2.

Table 4.2 Explainable properties of boosted rulesets

Properties	Values
Local or global	Global and local
Linear or non-linear	Non-linear (can be axis parallel)
Monotonic or non-monotonic	non-monotonic
Feature interactions captured	Yes
Model complexity	Medium to high

Fig. 4.1 Boosted rules with
depth of tree 1

Mined rules:

Glucose <= 123.5
BMI <= 26.45
Age <= 28.5 Glucose <= 166.5
BMI <= 26.45
BMI <= 30.95
DiabetesPedigreeFunction <= 0.21
Age <= 22.5
BloodPressure <= 53.0
Age <= 56.5
Glucose <= 99.5
DiabetesPedigreeFunction <= 0.71
DiabetesPedigreeFunction <= 1.39
Glucose <= 122.5
Glucose <= 99.5
BMI <= 48.10
Pregnancies <= 7.5

Mined rules:

Glucose <= 123.5 Age <= 28.5 BMI <= 30.05
Age <= 28.5 Glucose <= 165.5 DiabetesPedigreeFunction <= 0.21
BMI <= 26.45 Glucose <= 153.5 BloodPressure <= 53.0
BMI <= 26.45 Glucose <= 153.5 Glucose <= 91.5
BMI <= 22.80 BloodPressure <= 94.0 Insulin <= 186.5
Glucose <= 166.5BMI <= 45.4 Glucose <= 188.5
Age <= 22.5 SkinThickness <= 33.5 Age <= 62.5
BMI <= 22.80 BloodPressure <= 94.0 BMI <= 23.0
BMI <= 22.80 BloodPressure <= 94.0 DiabetesPedigreeFunction <= 1.4
DiabetesPedigreeFunction <= 0.71 DiabetesPedigreeFunction <= 0.69 BMI <= 26.45
DiabetesPedigreeFunction <= 1.4 DiabetesPedigreeFunction <= 1.17 Age <= 26.5
BloodPressure <= 74.5 BMI <= 22.75 Glucose <= 179.5
Insulin <= 48.5 SkinThickness <= 28.5 Insulin <= 63.5
DiabetesPedigreeFunction <= 1.4 DiabetesPedigreeFunction <= 1.17 Age <= 26.5
Age <= 42.5 BMI <= 26.45 Pregnancies <= 0.5
BMI <= 26.45 Glucose <= 106.5 Glucose <= 121.5
Glucose <= 144.5 BMI <= 26.45 DiabetesPedigreeFunction <= 0.54
Glucose <= 144.5 DiabetesPedigreeFunction <= 1.10
Insulin <= 542.5 BMI <= 48.10 BMI <= 45.95 Age <= 24.0

Fig. 4.2 Boosted rules with depth of tree 2

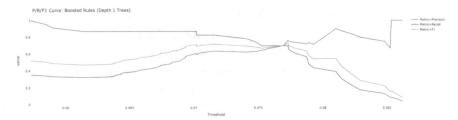

Fig. 4.3 Precision-recall curves for boosted rules with depth of tree 1

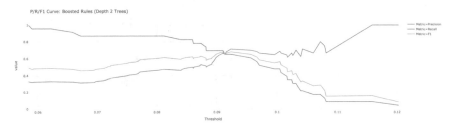

Fig. 4.4 Precision-recall curves for boosted rules with depth of tree 2

4.3.2 Explainable Boosting Machines (EBM)

Explainable Boosting Machine (EBM) is an interpretable algorithm developed by
Nori et al. to not sacrifice accuracy for interpretability [Nor+19]. EBM is an
extension of GAM where in one single iteration, a small tree is built from a single
feature, then through boosting the residuals are updated, and the next feature is used
to construct the tree; this is repeated for all the features [HT90b]. Thus, a round-
robin pass is done through all the features with a low learning rate to remove the
dependency on features' order. In the next iteration, another round-robin process is
started similarly and the process iterates for a large number of iterations (around
10,000).

$$g(\mathbb{E}(y|\mathbf{x})) = \mathbf{w}_0 + f_1(\mathbf{x}_1) + \cdots + f_d(\mathbf{x}_d) \tag{4.1}$$

In the end, as shown in Fig. 4.5, each feature $f_d(\mathbf{x}_d)$ is summarized as a function
from all the iterations. This function thus becomes the representation that gets
additively combined for all features to give the final model.

Interpretation of EBM: EBM outputs global explanation in terms of variable
importance for each feature, score output for each feature that explains the
odds and the evaluations like ROC curves. EBM also gives local explanation

(continued)

for an instance with predicted and actual values along with explanation on
which features contributed for the decision.

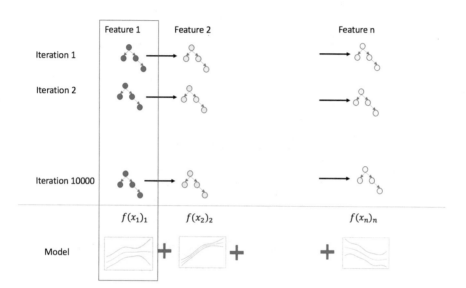

Fig. 4.5 Training process for EBM

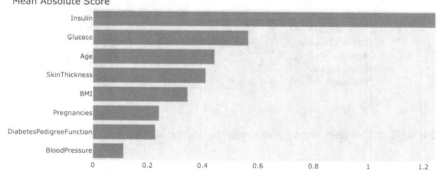

Fig. 4.6 Global variable importance

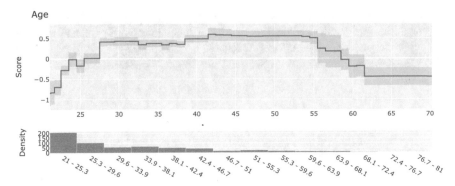

Fig. 4.7 Scores for the feature -age

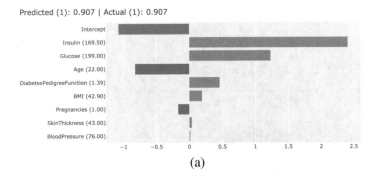

(a)

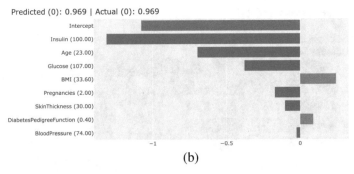

(b)

Fig. 4.8 Local predictions for a (**a**) positive and (**b**) negative instance

---

Observations:

- Figure 4.6 shows features *Insulin*, *Glucose*, and *Age* are the most discriminant features while *BloodPressure* is the least important one.

Table 4.3 Explainable properties of EBM

Properties	Values
Local or global	Global and local
Linear or non-linear	Non-linear
Monotonic or non-monotonic	Monotonic
Feature interactions captured	Yes
Model complexity	High

- Figure 4.7 shows the score output for *Age* that shows the pattern where at a younger age ($Age \leq 30$) the odds of being diabetic are low and then it increases, stays constant till middle age ($Age \leq 60$) and finally reduces back. Ability to visualize and interpret each feature score makes the method a complete whitebox one.
- The local outputs for the positive and the negative instances as shown in Fig 4.8a and b show which features contribute to classifying the instances.

Explainable properties of EBM are shown in Table 4.3.

4.3.3 RuleFit

RuleFit algorithm combines the accuracy of tree ensembles and a linear model's interpretability to give a robust, interpretable regression model [FP+08]. The structural form for the ensembles is given by

$$F(\mathbf{x}) = a_0 + \sum_{m=1}^{M} a_m f_m(\mathbf{x}) \tag{4.2}$$

where M is the size of the ensemble, $f_m(\mathbf{x})$ are the ensemble base learners, and $a_{m_0}^{M}$ are the linear weights used to combine them to form the ensemble. The base learners in the RuleFit algorithm are simple rules which take a conjunctive form given by

$$r_m(\mathbf{x}) = \prod_{j=1}^{n} I(x_j \in s_{jm}) \tag{4.3}$$

where S_j is the set of all possible values for the feature x_j, $x_j \in S_j$ and s_{jm} is the specified subset of those values, $s_{jm} \subset S_j$ and $I(\cdot)$ is the indicator that is true for the condition. For continuous features, subsets are taken to be contiguous intervals

$$s_{jm} = (t_{jm}, u_{jm}) \tag{4.4}$$

defined by the upper and the lower bound, $t_{jm} \leq x_{jm} \leq u_{jm}$. Total number of rules created from M ensemble with t_m terminal nodes is

$$K = \sum_{m=1}^{M} 2(t_m - 1) \tag{4.5}$$

The predictive model thus becomes

$$F(\mathbf{x}) = \hat{a}_0 + \sum_{k=1}^{K} \hat{a}_k r_k(\mathbf{x}) \tag{4.6}$$

and the weights can be solved by

$$\{\hat{a}_k\}_0^K = \underset{\{\hat{a}_k\}_0^K}{\arg \min} \sum_{i=1}^{N} L\left(y_i, a_0 + \sum_{k=1}^{K} a_k r_k(\mathbf{x}_i)\right) + \lambda \cdot \sum_{k=1}^{K} |a_k| \tag{4.7}$$

where the first term measures the prediction risk on the training sample and second one is the "lasso" regularization penalty for the coefficient of the base learners.

> Interpretation of RuleFit: RuleFit algorithm has similar interpretation to a linear model, except that the features can be decision rules with combinations connected through conjunctions instead of just the features. The interpretation will be—if the feature is true (since a decision rule is Boolean, it is when the rule applies) and increases by one unit, the predicted outcome changes by the corresponding feature weight. Most RuleFit algorithm implementations also output the relative importance of each decision rule as a feature and the support (fraction of the training data matching the rule) to further whitebox the algorithm.

	rule	type	coef	support	importance
713	bmi > 30.010000228881836 and bmi > 28.59749984741211 and smoker > 0.5	rule	12529.641445	0.112853	3964.536570
307	age <= 58.5 and smoker <= 0.5	rule	-2008.459390	0.742947	877.714900
811	bmi > 22.887499809265137 and smoker > 0.5	rule	2132.713404	0.141066	742.375026
386	children <= 3.5 and smoker <= 0.5	rule	-1571.686972	0.783699	647.097795
15	age <= 42.5 and smoker <= 0.5	rule	-1136.919132	0.426332	562.255731

Fig. 4.9 RuleFit algorithm output with decision rules, coefficients, importance, and support for the insurance claim regression data

> **Observations:**
>
> - Many features such as *region, sex*, etc. play no role in any decision rules in Fig. 4.9 indicating that their importance is relatively low in the dataset.
> - The first rule, *bmi* > 28.5 *AND smoker* > 0.5 *AND bmi* > 30 can be interpreted as *bmi* > 28.5 *AND smoker*! = 0, thus capturing the combination of obesity and smoking that is an indicator of the high *charge* due to largest positive coefficient (12529.64) and importance (3964.53).
> - The ability of the algorithms to combine features like *age* and *smoker* (*age* < 58.5 *AND smoker* < 0.5) with a negative coefficient (−2008.45) and a large support (0.74) explains for most of the lower charges of how age and not smoking helps in lowering the cost.
> - Rules such as *children* <= 3.5 *AND smoker* <= 0.5 with a negative coefficient (−1571.66) and a large support (0.78) give an explanation for most of the lower charges of how fewer children and not smoking helps in lowering the cost.

Explainable properties of RuleFit are shown in Table 4.4.

4.3.4 Skope-Rules

Skope-Rules are very similar to the RuleFit algorithm discussed before. Skope-Rules first create simple trees by fitting classification and regression trees to sub-samples [Gar+17]. Rules are then extracted from the tree ensembles and evaluated out of the bag, and only those above a certain precision are selected and merged. The rules are then further simplified by removing duplicates or those

Table 4.4 Explainable properties of RuleFit

Properties	Values
Local or global	Global and local
Linear or non-linear	Non-linear
Monotonic or non-monotonic	Non-monotonic
Feature interactions captured	Yes
Model complexity	Medium to high

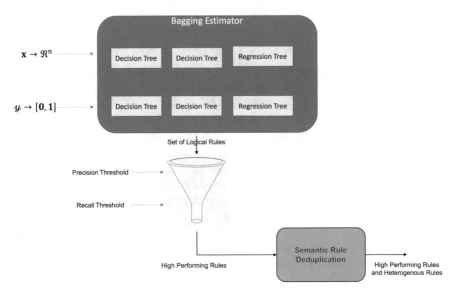

Fig. 4.10 How are Skope-Rules derived from the data?

that are too similar to others based on features, conditions, and on a similarity threshold of their supports. The weights in Skope-Rules are not optimized as an L1-regularized optimization problem, but weights are simply proportional to the rules' out-of-the-box affiliated precision. Figure 4.10 shows the entire process as a schematic.

> Interpretation of Skope-Rules:
>
> For training data $(x_i, y_i)_{i=1}^{n}$ where the $x_i \in X$ are the features, and y_i are the labels (binary), the Rules are of the form
> $constant ::= digit \mid digit\ constant$
> $digit ::= 0|1|2|3|4|5|6|7|8|9$
> $boolean\_operator ::= < \mid > \mid \leq \mid \geq$
> $variable ::= x_i\ where\ i \in 1, d$
> $boolean\_condition ::= variable\ boolean\_operator\ constant$
> $rule := boolean\_condition \mid rule\ AND\ boolean\_condition \implies y_i$

Figure 4.11 shows rules generated by running it on the diabetes training dataset.

Rule 1: *SkinThickness* > 16.0 and *Insulin* > 121.0 and *Age* > 27.5 diabetes (0.85, 0.70)

Rule 2: *Insulin* <= 187.5 and *Insulin* > 109.0 and *Age* > 26.5 diabetes (0.86, 0.63)

Rule 3: *Insulin* <= 169.75 and *Insulin* > 166.5 diabetes (0.98, 0.56)

Rule 4: *SkinThickness* > 31.5 and *Insulin* > 109.0 and *Age* <= 28.5 diabetes (0.68, 0.17)

Rule 5: *Insulin* <= 169.75 and *Insulin* > 124.0 and *Age* <= 28.5 diabetes (0.62, 0.14)

Rule 6: *Glucose* > 157.5 and *Insulin* > 169.75 diabetes (0.82, 0.13)

Fig. 4.11 Skope results on diabetes dataset

Table 4.5 Explainable
properties of Skope-Rules

Properties	Values
Local or global	Global and local
Linear or non-linear	Non-linear
Monotonic or non-monotonic	Non-monotonic
Feature interactions captured	Yes
Model complexity	Medium to high

Observations:

- Figure 4.11 shows six independent rules each with conditions on features and ordered by performance metrics of precision and recall associated with those rules estimated on the training data. The first rule has highest precision and recall and so on.
- Many features such as *Pregnancies*, *DiabetesPedigreeFunction*, and *BMI* play no role in any rules indicating that their importance is relatively low in the dataset for this model. Most rules are centered around *Insulin*, *Age*, *SkinThickness*, and *Glucose*, indicating their relative importance in the model.
- All the rules are generally easily interpretable by the experts. For example, the rule *SkinThickness* > 16.0 and *Insulin* > 121.0 and *Age* > 27.5 gives a very simple combination of three factors and their thresholds for classifying patients as diabetic with high precision and recall.

Explainable properties of Skopes-Rules are shown in Table 4.5.

4.3.5 Iterative Random Forests (iRF)

Random Forest (RF) is an ensemble technique that leverages the high-order interactions between the features to obtain the state-of-the-art performance. However, one of the biggest challenges with RF is in interpreting these interactions in the ensemble model. Random Forests also suffer from instability when there is a change in the training dataset; it generates trees with different splits and different decision paths. Iterative Random Forest (iRF) algorithm addresses most of these issues by building on the RFs that searches for stable, high-order interactions [Bas+18]. The iterative random forest algorithm sequentially develops feature-weighted RFs to perform soft dimension reduction of the feature space and stabilizes decision paths. Then the fitted RFs are decoded using a generalization of the Random Intersection Trees algorithm (RIT). This procedure enables finding high-order feature combinations that are common on the RF decision paths. Let us consider the binary classification setting with training data $\mathcal{D}$ in the form $\{(\mathbf{x}_i, yi)\}_{i=1}^{n}$ with continuous or categorical d features $\mathbf{x} = (x_1, ..., x_d)$, and a binary label $y \in 0, 1$. The entire algorithm can be summarized as

1. **Iteratively re-weighted RF**: Breiman's original RF assigns uniform weights to all the features $RF(1/p, ..., 1/p)$. The iRF algorithm iteratively grows K feature-weighted RFs $RF(w(k)), k = 1, ..., K$ on the data $\mathcal{D}$. The first iteration of iRF ($k = 1$) starts similar to RF with $w(1) := (1/d, ..., 1/d)$ and it stores the feature importance (mean decrease in Gini impurity) of the d features as $v^{(1)} = (v_1^{(1)}, ..., v_p^{(1)})$. For iterations $k = 2, ..., K$, the iRF builds a weighted RF with weights set equal to the RF feature importance from the previous iteration, i.e., $w^{(k)} = v^{(k-1)}$.

2. **Generalized RIT** The RIT process is designed to help find subsets $S \subset 1, ..., d$ of features that not only make the two classes separable but are also found more in one class $C \in 0, 1$ relative to the other. For each tree $t = 1, ..., T$ in the output tree ensemble of the last step of feature-weighted RF, all leaf nodes are collected and indexed as $j_t = 1, ..., J(t)$. Every feature-label pair $(\mathbf{x}_i, yi)$ has a representation to a tree t by (I_{it}, Z_{it}), where (I_{it} is the set of unique feature indices falling on the path of the leaf node containing $(\mathbf{x}_i, yi)$ in the tth tree. Hence, each $(\mathbf{x}_i, yi)$ produces T index set and label pairs corresponding to the T trees. Next step is to aggregate these pairs across all the training data and trees as

$$R = (I_{it}, Z_{it}) : \mathbf{x}_i \ falls \ in \ leaf \ node \ i_t \ of \ tree \ t \qquad (4.8)$$

Collection of higher order interactions as a subset S is done based on

$$P_n(S|Z = C) := \frac{\sum_{i=1}^{n} \mathbb{1}(S \subseteq I_i)}{\sum_{i=1}^{n} \mathbb{1}(S \subseteq C)} \qquad (4.9)$$

where P_n is the empirical probability distribution and $\mathbb{1}(\cdot)$ is the indicator function. For given thresholds $0 \leq \theta_0 < \theta_1 \leq 1$ RIT performs a randomized search for interactions S satisfying

$$P_n(S|Z = 1) \geq \theta_1, P_n(S|Z = 0) \leq \theta_0 \qquad (4.10)$$

3. **Bagged stability scores**: Another bagging process is carried out after finding the subset of recovered interactions S. Next is to generate bootstrap samples of the data $\mathcal{D}_{(b)}, b = 1, \ldots, B$, fit $RF^{(w(K))}$ on each bootstrap sample $\mathcal{D}_{(b)}$, and use the generalized RIT to identify interactions $S_{(b)}$ across each bootstrap sample. The stability score of an interaction $S \in \bigcup_{b=1}^{B} S_{(b)}$ is

$$stability(S) = \frac{1}{B} \sum_{b=1}^{B} \mathbb{1}\{S \in S_{(b)}\} \tag{4.11}$$

representing the proportion of times (out of B bootstrap samples) an interaction appears as an output of RIT.

Interpreting Iterative Random Forest: As shown in Figs. 4.12 and 4.13, the output will have multiple decision trees and every decision tree node has the splitting feature and threshold, splitting metric value and population in each class. The Gini score quantifies the purity of the node/leaf. A Gini score greater than zero implies that samples contained within that node belong to different classes. A Gini score of zero means that the node is pure, i.e., that node consists of representatives from only one class.

Observations:

- Figure 4.12 shows an interesting condition combining *Insulin* and *BMI* capturing a non-diabetic population (303 samples) with *Insulin* ≤ 121.0 *AND BMI* ≤ 50.0.
- Figure 4.13 shows another interesting condition combining *Glucose* and *Age* capturing a non-diabetic population (218 samples) with *Glucose* ≤ 144.5 *AND Age* ≤ 28.5 with low glucose at young age.
- Figures 4.12 and 4.13 show high glucose and high insulin impact on diabetes through conditions *Insulin* > 121.0 *AND Glucose* > 157.5 and *Glucose* > 162.5, respectively.

Explainable properties of Iterative Random Forests are shown in Table 4.6.

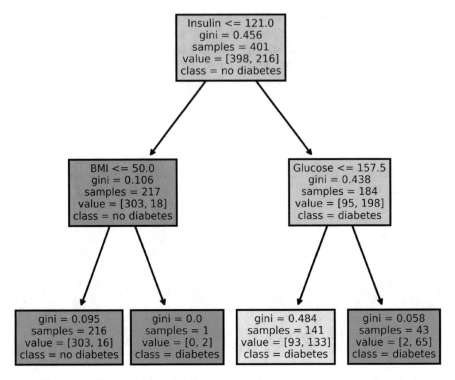

Fig. 4.12 A weighted decision tree from iRF on the diabetes dataset

Table 4.6 Explainable properties of Iterative Random Forest

Properties	Values
Local or global	Global and local
Linear or non-linear	Non-linear
Monotonic or non-monotonic	Non-monotonic
Feature interactions captured	Yes
Model complexity	Medium to high

4.4 Decision Tree-Based

Most decision tree methods such as CART, ID3, and C4.5 take a top-down approach in tree building. At each step of the partitioning process, the algorithms attempt to find the best split to partition the current data using some heuristics, such as minimizing the data's impurity through entropy metrics, as an example. This recursive process continues top-down till it achieves the stopping criteria and is known as the growth phase. Most algorithms then follow a pruning procedure bottom-up to remove branches that are redundant or weak in discrimination. This approach often leads to suboptimal trees that are locally optimal and have inefficiencies such as weak splits in the initial part of the tree that cannot be

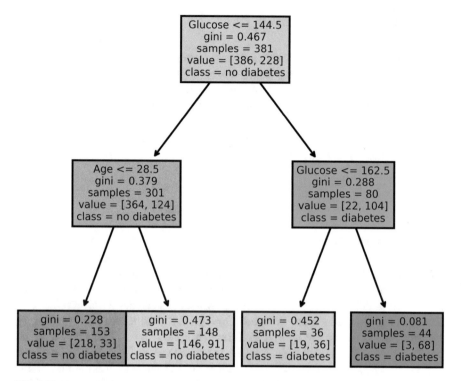

Fig. 4.13 A weighted decision tree from iRF on the diabetes dataset

undone. Finding optimal sparse decision trees is a NP hard problem as it leads to combinatorial explosion based on the number of features [Rud19a]. The efforts towards overcoming these issues were started by the machine learning community and statisticians. Kristin Bennett's seminal work was on global tree optimization using an iterative linear programming algorithm and variants to generate a globally optimum tree instead of a greedy approach [BB96]. Norouzi et al. took the direction of optimizing the convex-concave upper bound of the empirical loss using Stochastic Gradient Descent instead of optimizing the tree model's empirical loss itself and thus achieved a solution for the multiclass problem instead of just binary classification [Nor+15]. Many researchers also worked on improvements in the splitting criteria using various statistical techniques such contingency tables, hypothesis testing, removing biases, using unbiased variable selection, etc. [LV88, LS97, KL01, HHZ06, HS07]. Segal was among the first to extend decision tree algorithms like CART to longitudinal data by using as node impurity a function of the likelihood of an autoregressive model [Seg92]. Various techniques in avoiding covariance estimation and reducing the dimensionality by fitting trajectories to the spline curves were employed [YL99, LS04]. Sela and Simonoff introduced an expectation maximization based method which fits a model consisting of the sum of a random-effects term and a tree-structured term [SS12].

4.4.1 Optimal Classification Trees

Optimal Classification Trees (OCT) and their variants build a complete optimal tree given the depth constraint using Mixed-Integer Optimization (MIO) [BD17]. MIO helps in formulating an Optimal Decision Tree by finding answers to two critical questions: (1) Decisions: Which features to split on and which label to predict for the splits? (2) Outcomes: Which region a data instance ends up in and whether that instance is correctly classified. Given the training data $(\mathbf{x}_i, y_i), i = 1, \ldots, n$ the goal is to build a tree T such that

$$\min R_{xy} + \alpha|T|$$
$$s.t N_x(l) \geq N_{min} \forall l \in leaves(T) \tag{4.12}$$

where R_{xy} is classification error to minimize of the tree T, α is the complexity parameter, $|T|$ is the number of branch nodes in the tree T, and $N_x(l)$ is the is the number of training instances in the leaf node l. Note that this solves the tree optimally in one-shot as compared to recursive process of growing and pruning. Now to formulate the problem in terms of MIO, we consider a univariate tree as shown in Fig. 4.14 which has branches $\mathfrak{B} = \{1, 2, 3\}$ and leaves $\mathfrak{L} = \{4, 5, 6, 7\}$. Variables $\mathbf{a}_t, b_t$ define the split at each branch node $t \in \mathfrak{B}$. Elements of $\mathbf{a}_t$ are binary, i.e., $\mathbf{a}_t \in 0, 1$ and at a given time one has the value of 1, i.e., $\sum_{j=1}^{d} \mathbf{a}_{jt} = 1$, so that one feature is considered for the split parallel to the axis. Making the choices discrete, the data instance has to be assigned to the leaf l with a binary variable $z_{it} = 1$ if point i is assigned to leaf t, 0 otherwise such that $\sum_{t \in \mathfrak{L}} = 1$ to ensure each instance is assigned to a leaf. Enforcing the splitting rules,

$$\mathbf{a}_m^T \mathbf{x}_i + \epsilon \leq b_t + M_1(1 - z_{it}) \, i = 1, \ldots, n \, \forall m \in A_L(t) \tag{4.13}$$

$$\mathbf{a}_m^T \mathbf{x}_i \geq b_t + M_2(1 - z_{it}) \, i = 1, \ldots, n \, \forall m \in A_R(t) \tag{4.14}$$

where $A_L(t)$ and $A_R(t)$ are the left-branch and right-branch ancestors of t. MIO solvers cannot handle strict inequality like $<$ and hence a small a small constant ϵ is added to the left-hand side of Eq. 4.13. The largest possible value for $\mathbf{a}_t^T (\mathbf{x}_i + \epsilon) - b_t$ is $1 + \epsilon_{max}$ where $\epsilon_{max} = \max_j\{\epsilon_j\}$. Therefore M_1 can be set as $M_1 = 1 + \epsilon_{max}$ and since largest possible value of $b_t - \mathbf{a}_t^T \mathbf{x}_i$ is 1, M_2 can be set to 1. The objective is to minimize the misclassification error, so an incorrect label prediction has cost 1, and a correct label prediction has cost 0. Thus we can define

$$Y_{ik} = \begin{cases} +1, if \ y_i = k \\ -1, otherwise \end{cases} \quad k = 1, \ldots, K, i = 1, \ldots, n \tag{4.15}$$

For each leaf node, the best class to assign is the most common label among the instances assigned to that node. If N_{kt} is the number of instances of label k in node t, and N_t is the total number of instances in node t, then

$$N_{kt} = \frac{1}{2} \sum_{i=1}^{n} (1 + Y_{ik}) z_{it} \; \forall k = 1, \ldots, K, t \in T_L \tag{4.16}$$

$$N_t = \sum_{i=1}^{n} z_{it} \; \forall t \in T_L \tag{4.17}$$

The misclassification error in each node L_t is going to be equal to the number of instances in the node less the number of instances of the most common label and given by

$$L_t = N_t - \max_{k=1,\ldots,K} \{N_{kt}\} = \min_{k=1,\ldots,K} \{N_t - N_{kt}\} \tag{4.18}$$

This loss can be used in the Eq. 4.12 with MIO solves to get optimal trees in one-shot. Other enhancements are done to the base algorithms such as hyperplane splits instead of axis parallel splits, search optimizations, etc. to further improve the effectiveness.

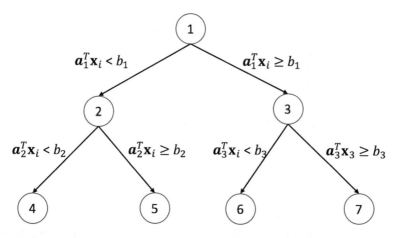

Fig. 4.14 Optimal classification trees

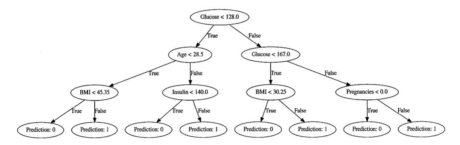

Fig. 4.15 Optimal classification tree for diabetes dataset

Table 4.7 Explainable properties of Optimal Classification Trees

Properties	Values
Local or global	Global and local
Linear or non-linear	Non-linear
Monotonic or non-monotonic	Non-monotonic
Feature interactions captured	Yes
Model complexity	Medium to high

Observations:
Figure 4.15 shows the optimal tree obtained on the diabetes training data. Rules such as $Glucose < 128\ AND\ Age < 28.5\ AND\ BMI > 45.35 \implies class = diabetes$ and $Glucose < 128\ AND\ Age < 28.5\ AND\ Insulin > 140 \implies class = diabetes$ clearly captures how high BMI or Insulin in young women can lead to diabetes. Also, the right tree with $Glucose > 167\ AND\ BMI > 30.25 \implies class = diabetes$ captures the general high glucose and obesity relationship with diabetes.

Explainable properties of Optimal Classification Trees are shown in Table 4.7.

4.4.2   Optimal Decision Trees

Decision tree optimization is notably complex from a computational perspective but crucial for interpretable machine learning. Recently, optimization breakthroughs have allowed practical algorithms to find optimal decision trees for a reasonable number of features. This section describes some of the breakthroughs and innovations resulting in optimal and interpretable decision trees.

4.4.2.1 Optimal Sparse Decision Trees

Optimal sparse decision tree (OSDT) is a decision tree algorithm equivalent to Certifiably Optimal RulE ListS (CORELS) [HRS19]. OSDT overcomes the optimality issues of greedy decision tree algorithms like CART and C4.5 by building a globally optimal tree using analytical bounds to narrow the search space, improved data structures and caching techniques. Let $\{(x_n, y_n)\}_{n=1}^{N}$ represent the training data, where $x_n \in \{0, 1\}^M$ are binary features and $y_n \in \{0, 1\}$ are the labels. Let $\mathbf{x} = \{x_n\}_{n=1}^{N}$ and $\mathbf{y} = \{y_n\}_{n=1}^{N}$ and thus $x_{n,m}$ denote the m-th feature of x_n. For a tree d, the optimization objective can be written as

$$R(d, \mathbf{x}, \mathbf{y}) = l(d, \mathbf{x}, \mathbf{y}) + \lambda H_d. \qquad (4.19)$$

where $R(d, \mathbf{x}, \mathbf{y})$ is the regularized empirical risk, $l(d, \mathbf{x}, \mathbf{y})$ is the misclassification error of d, H_d is the number of leaves in the tree d, and λ is the regularization term. A λ value of 0.01 indicates adding a penalty of 1% in misclassification by adding one extra leaf to the tree. The optimization process depends on various theorems that reduce the trees from growing and thus narrowing the search space. Next, we will give a high level overview of these theorems from their utility perspective in reducing the size of the trees. For the interested readers, the paper proves each of these in a formal way.

1. **Equivalent points bound**: We will classify at least the minority label of the data wrong for every set of equivalent points. If multiple instances have identical feature values but opposite labels, we know that any model will make an error.
2. **Hierarchical objective lower bound**: Lower bounds of parent tree holds for every child of that parent.
3. **One-step look-ahead lower bound**: Give the number of leaves, if a tree does not achieve the given accuracy, then all children of that tree can be pruned.
4. **Apriori bound on the number of leaves**: Every optimal decision tree has an a priori upper bound on the maximum number of leaves.
5. **Lower bound on node support**: For an optimal decision tree, the support traversing through each internal node must be at least 2λ
6. **Lower bound on incremental classification accuracy**: Each split of the node in the tree should result in a sufficient reduction in the loss. If the loss reduction is less than or equal to the regularization, then there exists at least one more split in the new child leaf nodes.
7. **Leaf permutation bound**: If there is a less accurate permutation of a leaf, it cannot be extended further.
8. **Leaf accuracy bound**: For each leaf in an optimal decision tree, the accuracy must be above a threshold.

Further, data structure enhancements such as storing the bounds, intermediate metrics of accuracy, labels of the leaf, etc. for the entire tree and the individual leaves facilitates the incremental computation of the lower bound and the objective. Using priority queues, where each entry is a tree and removing an entry results in more child trees, further enhances the incremental computation and speedup. Bit-vector of data instances as features described by leaves and permutation maps caching the structures gives the algorithm further edge in incremental computations.

Best Tree

~~~~~~ Leaf 1: Prediction = 0 ~~~~~~
Glucose=(79.6-99.5]

~~~~~~ Leaf 2: Prediction = 1 ~~~~~~
NOT *Glucose*=(79.6-99.5]
  && *Glucose*=(179.1-inf)

~~~~~~ Leaf 3: Prediction = 1 ~~~~~~
NOT '*Glucose*=(179.1-inf)
  && NOT '*Glucose*=(79.6-99.5]
  && *Glucose*=(159.2-179.1]

~~~~~~ Leaf 4: Prediction = 0 ~~~~~~
NOT *Glucose*=(159.2-179.1]
  && NOT *Glucose*=(179.1-inf)
  && NOT *Glucose*=(79.6-99.5]
  && *BMI*=(20.13-26.84]

~~~~~~ Leaf 5: Prediction = 0 ~~~~~~
NOT *Glucose*=(159.2-179.1]
  && NOT *Glucose*=(179.1-inf)
  && NOT *BMI*=(20.13-26.84]
  && NOT *Glucose*=(79.6-99.5]
  && *Age*=(-inf-27]

~~~~~~ Leaf 6: Prediction = 1 ~~~~~~
NOT *DiabetesPedigreeFunction*=(-inf-0.3122]
  && NOT *Glucose*=(159.2-179.1]
  && NOT *Glucose*=(179.1-inf)
  && NOT *BMI*=(20.13-26.84]
  && NOT *Glucose*=(79.6-99.5]
  && NOT *Age*=(-inf-27]

~~~~~~ Leaf 7: Prediction = 0 ~~~~~~
NOT *Glucose*=(159.2-179.1]
  && NOT *Glucose*=(179.1-inf)
  && NOT *BMI*=(20.13-26.84]
  && NOT *Glucose*=(79.6-99.5]
  && NOT *Age*=(-inf-27]
  && *DiabetesPedigreeFunction*=(-inf-0.3122]

Fig. 4.16   OSDT tree for diabetes dataset

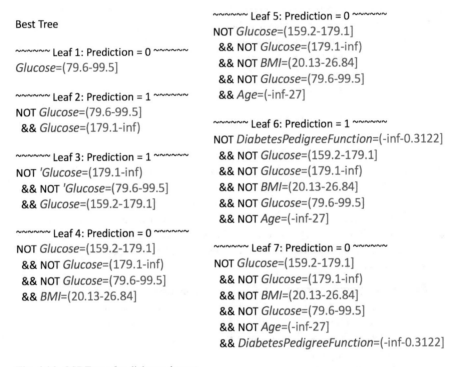

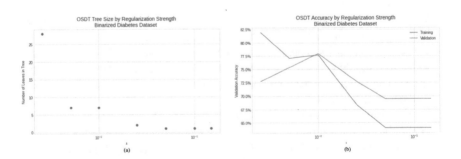

Fig. 4.17   Impact of regularization on leaf nodes and accuracy. (**a**) Leaf nodes. (**b**) Accuracy

Table 4.8 Explainable properties of optimal sparse decision trees

Properties	Values
Local or global	Global and local
Linear or non-linear	Non-linear
Monotonic or non-monotonic	Non-monotonic
Feature interactions captured	Yes
Model complexity	Medium to high

Observations:

- Since the inputs are itemset based, we discretized the diabetes data and then binarized it to have 0/1 values.
- Figure 4.17a and b show the correlation between number of leaf nodes in the tree, accuracy and regularization parameter λ. At $\lambda = 0.01$ the validation accuracy is highest and is around 77.5 and the tree has 7 leaf nodes as shown in Fig. 4.16.
- Figure 4.16 is the output of OSDT with 7 leaf nodes and many of the nodes have domain-specific interpretation. For example, the leaf node 5 which predicts non-diabetes excludes high glucose (>159.2), young age (<27) AND healthy BMI ($\geq 20.13 and \leq 26.84$). Adding just *DiabetesPedigreeFunction* in any range except low then becomes an indicator of diabetes, especially in young people as given in the leaf node 6.

Explainable properties of Optimal Sparse Decision Trees are shown in Table 4.8.

4.4.2.2 DL8.5

DL8.5 algorithm is another decision tree learning algorithm using itemset rules and branch and bound techniques to generate optimal decision trees [ANS20]. All of the following concepts contribute to the novelty of DL8.5 in generating optimal decision trees.

- The use of itemset mining to represent possible paths.
- Using maximum depth to limit as soon as itemsets go above a threshold and not using an attribute for branching below minimum support.
- The use of the cache to store intermediate results (including partial trees).
- The use of upper bound on quality to recursively find branches or prune the space if cannot reach the quality.
- The ability to use heuristics during the search such as information gain of the feature to during the search to increase the speed of finding good trees.

Interpretation of DL8.5: DL8.5 generates a tree of given length with each node being a binary feature (0/1) branching to left and right recursively till the leaf node is reached with class label and misclassification errors.

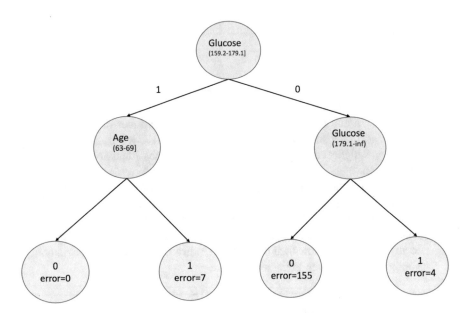

Fig. 4.18 DL8.5 with depth 2

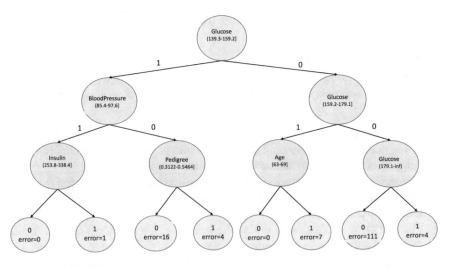

Fig. 4.19 Dl8.5 with depth 3

Observations:

- Continuous features are discretized using Fayyad and Irani MDL method and then binarized so that each feature is 0/1.
- Figures 4.18 and 4.19 are two trees with depth constraints of 2 and 3, respectively.
- When the size of the tree is increased from 2 to 3, the whole tree of size 2 is seen as a sub-tree of size 3, indicating the robust nature of the splits and tree.
- From classification performance, the testing error slightly increases from the tree of size 2–3 but the training error reduces indicating overfitting with size 3.
- The trees are completely interpretable and can generate rules to debug or whitebox the models, for example, Fig. 4.18 a rule can be extracted like (1) *If* $159.2 < Glucose \leq 179.1\ AND\ 63 < AGE \leq 69 \implies diabetes$

Explainable properties of DL8.5 are shown in Table 4.9.

Table 4.9 Explainable properties of DL8.5

Properties	Values
Local or global	Global and local
Linear or non-linear	Non-linear
Monotonic or non-monotonic	Non-monotonic
Feature interactions captured	Yes
Model complexity	Medium to high

4.4.2.3 Generalized and Scalable Optimal Sparse Decision Trees (GOSDT)

The GOSDT algorithm improved over previous OSDT with the following main changes (1) optimal decision trees cover various objectives such as F-score, AUC, and partial area under the ROC convex hull (2) training data with both continuous features along with binary features supported (3) novel dynamic programming with bounds algorithm to enable parallelism and reduce the run-time and memory usage of decision trees during the optimization/search. Let $\{(x_n, y_n)\}_{n=1}^{N}$ represent the training data, where $x_n \in \{0, 1\}^M$ are binary features and $y_n \in \{0, 1\}$ are the labels. Let $\mathbf{x} = \{x_n\}_{n=1}^{N}$ and $\mathbf{y} = \{y_n\}_{n=1}^{N}$ and thus $x_{n,m}$ denote the m-th feature of x_n. A tree d has leaf set $d = (l_1, \ldots, l_{H_d})$ containing H_d distinct leaves, where l_i is the classification rule of the leaf i and $\hat{y}_i^{leaf}$ is the label prediction for all data in leaf i. For a tree d, the optimization objective can be written as

$$R(d, \mathbf{x}, \mathbf{y}) = l(d, \mathbf{x}, \mathbf{y}) + \lambda H_d \qquad (4.20)$$

where $R(d, \mathbf{x}, \mathbf{y})$ is the regularized empirical risk, $l(d, \mathbf{x}, \mathbf{y})$ is the loss function of d, H_d is the number of leaves in the tree d, and λ is the regularization term. A λ value of 0.01 indicates adding a penalty of 1% in misclassification by adding one extra leaf to the tree. The loss function $l(d, \mathbf{x}, \mathbf{y})$ can be monotonic such as accuracy-based or rank statistics such as area under curve.

Some OSDT algorithm bounds have extensions to the objectives in GOSDT, particularly the **Upper Bound on Number of Leaves** and **Leaf Permutation Bound**. GOSDT introduces different bounds such as the **Hierarchical Objective Lower Bound, Incremental Progress Bound to Determine Splitting, Lower Bound on Incremental Progress, Equivalent Points Bound, Similar Support Bound, Incremental Similar Support Bound**, and a **Subset Bound** to search efficiently by reducing the search space. Dynamic programming starts with a full dataset and a naive label, iteratively splitting the dataset using each feature, consolidating the duplicates, and stopping when subsets cannot be split further. Each subset decides the best feature to split at, and at the end the optimal tree emerges as a DAG of the best features through the subsets.

> Interpretation of GOSDT: GOSDT generates a *if-then-else* patterns containing literals with features and values in conjunctions. Each pattern is an independent rule that results in a positive or negative class.

```
if 144 <= Glucose then:
    predicted class: 1
    misclassification penalty: 0.065
    complexity penalty: 0.025

else if Glucose < 144 then:
    predicted class: 0
    misclassification penalty: 0.191
    complexity penalty: 0.025
```

Fig. 4.20 GOSDT

Table 4.10 Explainable
properties of GOSDT

Properties	Values
Local or global	Global and local
Linear or non-linear	Non-linear
Monotonic or non-monotonic	Non-monotonic
Feature interactions captured	Yes
Model complexity	Medium to high

Observations:

- Continuous features are discretized using Fayyad and Irani MDL method. This results in around 79 categorical features.
- Figure 4.20 shows only one rule that captures range of feature *Glucose* (≥ 144) for diabetes and if not then no diabetes is quite interpretable. However, interactions with other features and more complex rules are not seen through this technique.
  Note: This rule took 1 h to run with hardware of 32 cores and 208 GB RAM with "precision limit" set high.

Explainable properties of GOSDT are shown in Table 4.10.

4.5 Rule-Based Techniques

A general inductive rule learning algorithm aims to learn a ruleset from a given training data covering the positive class. Most algorithms differ in how they learn individual rules; most of them employ a separate-and-conquer or covering strategy for combining rules into a rule set [Mic83b, Für99, Coh95b]. We can divide the separate-and-conquer rule learning into two main steps: First, create a single rule from the training data (the conquer step) and then remove all the examples covered by that rule (the separate step). The process is iterative and finishes only when there are no more positive instances left. This process ensures that every positive instance is covered at least by one rule (completeness), and no negative instance is included (consistency).

Various pruning heuristics or a stopping criterion have been introduced to improve the accuracy and speed up the performance [CB91, Coh95b, Für99]. The search space of decision trees of a given depth is much larger than the search space of rule lists of that same depth [Lar+18]. Exploring the search space to build optimal trees or rules using Bayesian techniques is common in decision tree methods and rule-based methods [DMS98, CGM02, CGM+10, Let+15]. Disjunctive normal form (DNF) models using tight bounds to guide the search space in rule induction are common in the literature [FW98].

4.5.1   Bayesian Or's of And's (BOA)

Bayesian ors of ands (BOA) is a classifier built with association rule patterns to cover positive classes maximally from the training data while keeping the model interpretable [Wan+15]. Let $\{\mathbf{x}_n, y_n\}$ represent the data, where $\mathbf{x}_n$ is a data vector of continuous or categorical features and $y_n \in \{0, 1\}$ is the label. A literal is a feature-value pair (e.g., x1='blue'), denoted as r. A pattern is a conjunction of literals (e.g., x1='blue' AND x2='<5'), denoted as a. Thus, rules have a form $a = r_1 \wedge \dots, r_n$ where $\wedge$ denotes the AND operation. A pattern set is a disjunction of patterns, denoted as A, and has the form $A = a_1 \vee \dots, a_m$, where $\vee$ denotes the OR operation. Consider a boolean function $h(\mathbf{x}_n, a)$ that evaluates if pattern a applies to the instance $\mathbf{x}_n$, then a classifier built from the pattern set A can be define as f_A

$$f_A(\mathbf{x}_n) = \begin{cases} 1 & \exists a \in A, h(\mathbf{x}_n, a) = 1 \\ 0 & \text{otherwise} \end{cases} \tag{4.21}$$

As shown in Fig. 4.21, the BOA methodology entails finding patterns that maximize the positive class coverage with minimal negative class coverage. BOA employs generative approach to the construction of these patterns set using two probabilistic approaches, viz. one with a Beta-Binomial prior, and the other with Poisson priors. In the Beta-Binomial model, the pattern length L is pre-determined, and the model uses L beta priors to control the probabilities of selecting patterns. In a Poisson model, the "shape" of a pattern set, which includes the number of patterns and lengths of patterns, is decided by drawing from Poisson distributions parameterized with user-defined values. The generative process is about filling in with literals by first randomly selecting the features and then randomly selecting values corresponding to each feature. Inference in the BOA model has computational challenges as it involves a search over exponentially many possible sets of patterns. Since each pattern is a conjunction of literals, the number of patterns increases exponentially with number of literals and size of the set of the patterns. Stochastic local search using simulated annealing that either adds/removes/changes literals or the patterns is used as the heuristic in building the classifier.

Interpretation of BOA: BOA generates a pattern set, with each pattern containing literals with features and values in conjunctions. Each pattern is an independent rule that results in a positive class.

Table 4.11 Explainable
properties of BOA

Properties	Values
Local or global	Global and local
Linear or non-linear	Non-linear
Monotonic or non-monotonic	Non-monotonic
Feature interactions captured	Yes
Model complexity	Medium to high

Observations:

- Continuous features are discretized using Fayyad and Irani MDL method and then binarized so that each feature is 0/1. This results in around 79 binary features.
- In Fig. 4.22, the first rule captures interaction between age (≥ 27), BMI range (20.13–26.84) and glucose range (79.6–99.5) for diabetes and is quite interpretable.
- The second rule captures various intervals of single feature, i.e., *Glucose* whose absence determines the class diabetes.

Explainable properties of BOA are shown in Table 4.11.

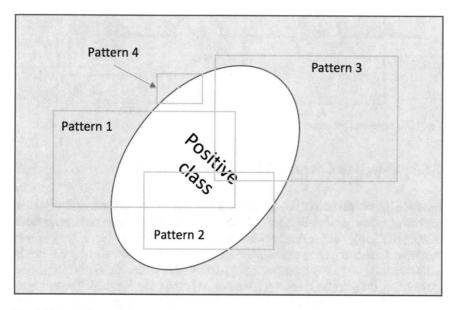

Fig. 4.21 Bayesian ors of ands

The patterns are mined! Time taken: 0.0103 min
The patterns are screened! Time taken: 0.237 min
Cooling rate is: 10
Progress: |=====================| Iteration 99/100 ME: 0.28 Score: 820.87

Rules
1. {$X.Age$-inf-27=0} AND {$X.BMI$ 20.13-26.84=0} AND {$X.Glucose$ 79.6- 99.5=0}
2. {$X. Glucose$ 119.4-139.3=0} AND   {$X. Glucose$ 99.5-119.4.=0}   AND   {$X.Glucose$ 79.6-99.5 =0}

Fig. 4.22 BOA results on diabetes dataset

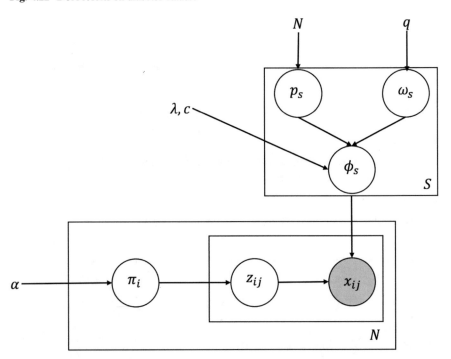

Fig. 4.23 Bayesian case model

4.5.2   Bayesian Case Model

Bayesian Case Model (BCM) provides a general framework using Case-based
reasoning, prototype classification, and clustering to give accurate and interpretable
models [KRS14]. Let the training data be represented as $\mathbf{x} = \{x_1, \ldots, x_N\}$ for N
instances. Each instance x_i can be represented as a random mixture over clusters in
a discrete mixture model representation. Assuming there are S clusters, an a priori
assumption, the graphical model representation is shown in Fig. 4.23. The vector π_i
represents the mixture weights over the S clusters for the ith instance x_i, $\pi_i \in \mathbb{R}^S$.
Each instance has P features, and we denote the jth feature of the ith observation
as x_{ij}. Each feature j of the instance x_i is associated with one of the clusters, the

index of the cluster for x_{ij} is given by z_{ij} and the full set of cluster assignments for instance-feature pairs is denoted by $\mathbf{z}$. Each z_{ij} takes on the value of a cluster index between 1 and S. The mixture weights π_i are generated according to a Dirichlet distribution, parameterized by hyperparameter α.

$$\pi_i \sim Dirichlet(\alpha) \; \forall i \qquad (4.22)$$

The cluster index z_{ij} is obtained for each x_{ij}, by sampling from a multinomial distribution with parameters π_i as

$$z_{ij} \sim Multinomial(\pi_i) \; \forall i, j \qquad (4.23)$$

Each feature for an instance x_{ij} is sampled from the feature distribution of the assigned subspace cluster $(\phi_{z_{ij}})$

$$x_{ij} \sim Multinomial(\phi_{z_{ij}}) \; \forall i, j \qquad (4.24)$$

The interpretability of BCM comes from the prototype-based cluster characterization instead of some predefined parametric distribution assumption in the standard mixture models. BCM identifies each cluster by a prototype, p_s, and a subspace feature indicator, ω_s. The prototype p_s for cluster s is defined as one instance in $\mathbf{x}$ that maximizes $p(p_s | \omega_s, \mathbf{z}, \mathbf{x})$. The prototype p_s is chosen randomly using a uniform distribution

$$p_s \sim Uniform(1, N) \; \forall s \qquad (4.25)$$

The subspace feature indicator given by $\omega_s \in \{0, 1\}^P$ can be seen is as an indicator that "turns on" the important features for characterizing cluster s and selecting the prototype p_s. The feature indicator (ω_{sj}) is generated according to a Bernoulli distribution with the hyperparameter q.

$$\omega_{sj} \sim Bernoulli(q) \; \forall s, j \qquad (4.26)$$

The ϕ_s is a data structure where each row ϕ_{sj} is a discrete probability distribution of possible outcomes for feature j. Let us consider θ as a vector of the possible outcomes of feature j (e.g., for feature 'gender', $\theta = [male, female]$), where θ represents a particular outcome for that feature (e.g., $\theta_v = male$). ϕ_{sj} generates in a way that it mostly takes outcomes from the prototype p_s for the cluster's important features. The g function characterizes—(1) when the feature j of cluster s is an important feature ($w_{sj} = 1$) and (2) when the value of the feature is identical to the value of the prototype of cluster s given by ($p_{sj} = \theta_v$) then (c, λ) the constant hyperparameters indicate how much copy of the prototype will be done in order to generate the instances.

$$g_{p_{sj},\omega_{sj},\lambda}(v) = \lambda(1 + c\mathbb{1}_{[\omega_{sj}=1 \ and \ p_{sj}=\theta_v]}) \tag{4.27}$$

The distribution of feature outcomes will be determined by g through

$$\phi_{sj} \sim Dirichlet(g_{p_{sj}}, \omega_{sj}, \lambda) \tag{4.28}$$

BCM uses collapsed Gibbs sampling for inference.

> **Interpretation of BCM:** BCM can be interpreted using the prototypes selected, the feature space representation, and the classifier built from the π matrix. Heatmap of features selected in the prototypes is one way of highlighting the importance of the features in the subspace. Visualizing the top features in subspace by deviation from global averages can help provide some interesting insight to the subspace groups. Finally, classifiers such as decision trees or other interpretable models using $\pi n \times S$ matrix where S is used as features give more interpretability at the classifier level.

> **Observations:**
>
> - Continuous features are discretized using Fayyad and Irani MDL method and then binarized so that each feature is 0/1. This results in around 79 binary features.
> - Figure 4.24 is the heatmap for all the binarized features in the 7 prototype subspaces and Fig. 4.25 gives information on each prototype based on top features in that subspace.
> - Figure 4.26 provides a decision tree view using π matrix.

Explainable properties of BCM are shown in Table 4.12.

4.5.3   Certifiably Optimal RulE ListS (CORELS)

Certifiably Optimal RulE ListS (CORELS) produces rule lists that are not only several times faster but guarantee optimality over the input categorical feature space. It leverages several efficiencies such as algorithmic bounds, effective data structures, and computational reuse to reduce the search space and increase the training speed [Lar+18]. Let $\{(x_n, y_n)\}_{n=1}^{N}$ represent the training data, where $x_n \in \{0, 1\}^J$ are binary features and $y_n \in \{0, 1\}$ are the labels. Let $\mathbf{x} = \{x_n\}_{n=1}^{N}$ and $\mathbf{y} = \{y_n\}_{n=1}^{N}$ and thus $x_{n,j}$ denote the j-th feature of x_n. A general representation of rule list is given

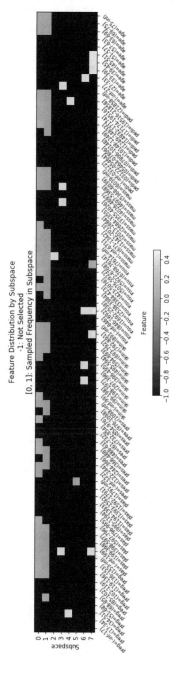

Fig. 4.24 Feature distribution by subspaces using BCM on binarized diabetes dataset

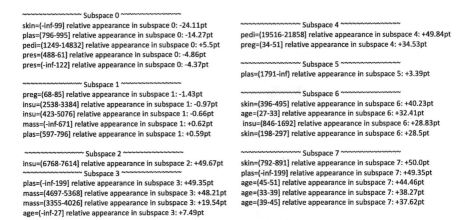

~~~~~~~~~~~~~~ Subspace 0 ~~~~~~~~~~~~~~
skin=(-inf-99] relative appearance in subspace 0: -24.11pt
plas=(796-995] relative appearance in subspace 0: -14.27pt
pedi=(1249-14832] relative appearance in subspace 0: +5.5pt
pres=(488-61] relative appearance in subspace 0: -4.86pt
pres=(-inf-122] relative appearance in subspace 0: -4.37pt

~~~~~~~~~~~~~~ Subspace 1 ~~~~~~~~~~~~~~
preg=(68-85] relative appearance in subspace 1: -1.43pt
insu=(2538-3384] relative appearance in subspace 1: -0.97pt
insu=(423-5076] relative appearance in subspace 1: -0.66pt
mass=(-inf-671] relative appearance in subspace 1: +0.62pt
plas=(597-796] relative appearance in subspace 1: +0.59pt

~~~~~~~~~~~~~~ Subspace 2 ~~~~~~~~~~~~~~
insu=(6768-7614] relative appearance in subspace 2: +49.67pt
~~~~~~~~~~~~~~ Subspace 3 ~~~~~~~~~~~~~~
plas=(-inf-199] relative appearance in subspace 3: +49.35pt
mass=(4697-5368] relative appearance in subspace 3: +48.21pt
mass=(3355-4026] relative appearance in subspace 3: +19.54pt
age=(-inf-27] relative appearance in subspace 3: +7.49pt

~~~~~~~~~~~~~~ Subspace 4 ~~~~~~~~~~~~~~
pedi=(19516-21858] relative appearance in subspace 4: +49.84pt
preg=(34-51] relative appearance in subspace 4: +34.53pt

~~~~~~~~~~~~~~ Subspace 5 ~~~~~~~~~~~~~~
plas=(1791-inf) relative appearance in subspace 5: +3.39pt

~~~~~~~~~~~~~~ Subspace 6 ~~~~~~~~~~~~~~
skin=(396-495] relative appearance in subspace 6: +40.23pt
age=(27-33] relative appearance in subspace 6: +32.41pt
insu=(846-1692] relative appearance in subspace 6: +28.83pt
skin=(198-297] relative appearance in subspace 6: +28.5pt

~~~~~~~~~~~~~~ Subspace 7 ~~~~~~~~~~~~~~
skin=(792-891] relative appearance in subspace 7: +50.0pt
plas=(-inf-199] relative appearance in subspace 7: +49.35pt
age=(45-51] relative appearance in subspace 7: +44.46pt
age=(33-39] relative appearance in subspace 7: +38.27pt
age=(39-45] relative appearance in subspace 7: +37.62pt

Fig. 4.25 Prototypes and their feature representation with BCM on binarized diabetes dataset

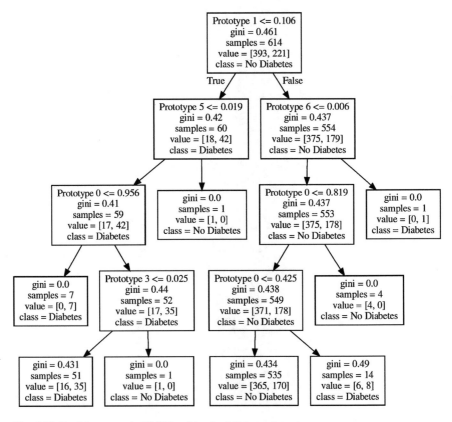

Fig. 4.26 Decision tree using BCM on binarized diabetes dataset

Table 4.12 Explainable
properties of BCM

Properties	Values
Local or global	Global and local
Linear or non-linear	Non-linear
Monotonic or non-monotonic	Non-monotonic
Feature interactions captured	Yes
Model complexity	Medium to high

by $d = (r_1, r_2, \ldots, r_K, r_0)$ where $K \geq 0$ is the length of the list. Thus the rule list consists of $(K+1)$-tuple with K distinct association rules given by $r_k = p_k \longrightarrow q_k$, for $k = 1, \ldots, K$ followed by a default rule r_0. The paper introduces alternate rule list representation given by $d = (d_p, \delta_p, q_0, K)$, where $d_p = (p_1, \ldots, p_K)$ are d's prefix, $\delta_p = (q_1, \ldots, q_K) \in 0, 1^K$ gives the label predictions associated with d_p, and $q_0 \in 0, 1$ is the default label prediction. The objective function that the algorithm optimizes for a rule list $d = (d_p, \delta_p, q_0, K)$ can be written as

$$R(d, \mathbf{x}, \mathbf{y}) = l(d, \mathbf{x}, \mathbf{y}) + \lambda K \tag{4.29}$$

where $l(d, \mathbf{x}, \mathbf{y})$ is the loss function measuring misclassification error and λ is the regularization parameter to control the size of the tree. A λ value of 0.01 indicates adding a penalty of 1% in misclassification by adding one association rule to the list. The entire optimization process depends on various theorems that reduce the trees from growing, reducing the search space and thus making the problem tractable. Next, we will give state these theorems from their utility perspective in reducing the size of the trees.

1. **Support bound**: If a rule's support is less than λ, that rule cannot be in the optimal rule list.
2. **Accuracy bound**: If a rule in the list does not classify at least λ fraction of data instances in the training set, that rule cannot be in an optimal rule list.
3. **Length bound**: The length of an optimal rule list is bounded by a function of λ, in short, rule lists cannot be too long.
4. **One-step look-ahead bound**: If a prefix's lower bound is within λ of the best current value of the objective, adding any rules to the list will add to a non-optimal rule list.
5. **Equivalent points bound**: We will classify at least the minority label of the data wrong for every set of equivalent points. If multiple instances have identical feature values but opposite labels, we know that any model will make an error.
6. **Permutation bound**: Only an optimal permutation of a set of rules can be extended to form an optimal rule list. If there exists a less accurate permutation of a rule, then it cannot be extended further.

CORELS performs systematized computations of these lower bound calculations and the objective values by efficient and within the context using an incremental branch and bound algorithm. Incremental computations using prefix trees (tries), pruning using a symmetry-aware map, and a priority queue to explore the search space further give the algorithm performance improvements.

Algorithm 2: CORELS

Data: Objective $R(d, \mathbf{x}, \mathbf{y})$, lower_bound $b(d_p, \mathbf{x}, \mathbf{y})$, set of antecedents $S = \{s_m\}_{m=1}^{M}$, training data $(\mathbf{x}, \mathbf{y}) = \{(_n, y_n)\}_{n=1}^{N}$ and regularization parameter λ

Result: Provably optimal rule list d^* with minimum objective R^*

Mine all rules with sufficient support;

Start with *rule_list* of size 1;

initialize the priority queue;

while *queue of rule lists is not empty* **do**

    for *each of the children of current prefix* **do**

        //test each of the theorems for lower bounds;

        current_best, current_objective= checkSupportBound();

        current_best,current_objective= checkAccuracyBound();

        current_best,current_objective= check(LengthBound();

        current_best,current_objective= checkOneStepLooakaheadBound();

        current_best,current_objective= checkEquivalentPointsBound();

        current_best,current_objective= checkPermutationBound();

    //check for bounds;

    if *lower_bound > current_best* **then**

        discard(prefix);

    else

        addToQueue();

    // check for objective;

    if *current_objective < current_best_objective* **then**

        update(*rule_list*);

    else

        discard();

return *rule_list*

Interpretation of CORELS: CORELS generates optimal rules with **If, elseIf** and **else** with antecedents having features with condition(s) equal $=$ or not equals! $=$ with binary $(0, 1)$ values.

```
Generated 86 rules
Mining done after 0.004888 seconds
Minority bound generation done after 0.002421 seconds

87 rules 768 samples

RULE 0: ( default ), support=768, card=0:
RULE 1: ( {'Pregnancies=\'(-inf-1.7]\''} ), support=246, card=1:
RULE 2: ( {'Pregnancies=\'(1.7-3.4]\''} ), support=178, card=1:
RULE 3: ( {'Pregnancies=\'(3.4-5.1]\''} ), support=125, card=1:
RULE 4: ( {'Pregnancies=\'(5.1-6.8]\''} ), support=50, card=1:
...
RULE 85: ( {'Age=\'(51-57]\''-not} ) support=737, card=1:
RULE 86: ( {'Age=\'(57-63]\''-not} ), support=743, card=1:
Labels (2) for 768 samples

RULE 0: ( {label=0} ), support=500, card=1:
RULE 1: ( {label=1} ), support=268, card=1:
LEARNING RULE LIST via OBJECTIVE Captured Symmetry Map
min(objective): 0.34896 -> 0.33469, length: 1, cache size: 12
before garbage_collect. num_nodes: 52, log10(remaining): 1
after garbage_collect. num_nodes: 52, log10(remaining): 1
min(objective): 0.33469 -> 0.32302, length: 2, cache size: 62
before garbage_collect. num_nodes: 97, log10(remaining): 1
after garbage_collect. num_nodes: 97, log10(remaining): 1
minimum objective: 0.3230208333
minimum lower bound in queue: 1.0000000000
final num_nodes: 1455
final num_evaluated: 11991
final min_objcctive: 0.32302
final accuracy: 0.73698

OPTIMAL RULE LIST
if ({'Glucose=\'(159.2-179.1]\''}) then ({label=1})
else if ({'Glucose=\'(179.1-inf)\''}) then ({label=1})
else ({label=0})
```

Fig. 4.27 Output of CORELS on diabetes dataset

Table 4.13 Explainable
properties of CORELS

Properties	Values
Local or global	Global and local
Linear or non-linear	Non-linear
Monotonic or non-monotonic	Non-monotonic
Feature interactions captured	Yes
Model complexity	Medium to high

Observations:

- Continuous features are discretized using Fayyad and Irani MDL method and then binarized so that each feature is 0/1.
- There are 86 rules mined initially and then the optimal rule are simple conditions on high **Glucose** ranges (Fig. 4.27).

Explainable properties of CORELS are shown in Table 4.13.

4.5.4   Bayesian Rule Lists

Bayesian Rule Lists (BRL) is a classification algorithm based on association rule mining and sampling to build compact decision list that is interpretable [Let+15]. Association rule mining has two steps—(a) finding frequent itemsets from the data and (b) generating strong association rules from the frequent itemsets. Many techniques generate frequent itemsets, such as Apriori and Frequent-Pattern Growth (FP-Growth). BRL uses FP-Growth. Finding the frequent pattern through the association rule mining is first finding the **support** for the pattern, given by

$$Support(x_j = value) = \frac{1}{n} \sum_{i=1}^{n} I(x_j^{(i)} = value) \qquad (4.30)$$

where x_j is the feature having a categorical value $value$ and I is the indicator function that returns 1 when the instance i has the feature $x_j = value$. Continuous features are first converted to categorical using algorithms such as Fayyad and Irani's MDL method. Based on the thresholds on the minimum support that user defines, many patterns can be discarded. BRL algorithm intends to find lists with few rules and short conditions. BRL approaches this goal by defining a distribution of decision lists with prior distributions for the length of conditions (preference for shorter rules) and the number of rules (preference for a shorter list). BRL uses posterior distribution to achieve this goal and is given by

$$\underbrace{p(d|x, y, A, \alpha, \lambda, \eta)}_{\text{posterior}} \propto \underbrace{p(y|x, d, \alpha)}_{\text{likelihood}} \cdot \underbrace{p(d|A, \lambda, \eta)}_{\text{prior}} \qquad (4.31)$$

Where d is a decision list, x are the features, y is the label, A is the set of pre-mined conditions, λ is the prior expected length of the decision lists, η is the prior expected number of conditions in a rule, and α is the prior pseudo-count for the positive and negative classes. The $p(d|x, y, A, \alpha, \lambda, \eta)$ is the posterior distribution on the decision list given the data, the labels, and the constraints. The $p(y|x, d, \alpha)$ is the likelihood of the observed y labels, given the data, decision list, and the prior assumptions. One of the assumptions in BRL is that the labels y are generated through the Dirichlet-Multinomial distribution. The $p(d|A, \lambda, \eta)$ quantifies the prior distribution of the decision lists given the parameters (λ, η) which are truncated Poisson distributions. Estimation of the statistics is done by modifying the initial random generated list to create more samples using the Markov chain Monte Carlo (MCMC) method.

Interpretation of BRL:
   If a training data $(x_i, y_i)_{i=1}^{n}$ where the $x_i \in X$ are the features, and y_i are the labels (binary), a Bayesian decision list has the following form
   if x obeys a_1 **then** y $\sim Binomial(\theta_1), \theta_1 \sim Beta(\alpha_1 + N_1)$
   else if x obeys a_2 **then** y $\sim Binomial(\theta_2), \theta_2 \sim Beta(\alpha_2 + N_2)$
   $\vdots$

   else if x obeys a_m **then** y $\sim Binomial(\theta_m), \theta_m \sim Beta(\alpha_m + N_m)$
   else y $\sim Binomial(\theta_0), \theta_0 \sim Beta(\alpha_0 + N_0)$
   The antecedents $a_{j=1}^{m}$ are the boolean conditions on the features; the vector $\alpha = [\alpha_1, \alpha_0]$ has a prior parameter for each of the two labels, and values α_1 and α_0 are prior parameters.
   N_j is the vector of counts, where $N_{j,l}$ is the number of observations x_i that satisfy condition a_j but none of the previous conditions $a_1, \dots, a_{j-1}$ and label $y_{i=l}$, 1 is either 1 or 0 in binary classification.

IF *Glucose* 154.5_to_inf **THEN** probability of diabetes: 80.0% (71.7%-87.2%)

ELSE IF *BMI* -inf_to_26.45 **THEN** probability of diabetes: 3.2% (0.9%-6.9%)

ELSE IF *Glucose* 123.5_to_154.5 **THEN** probability of diabetes: 54.3% (46.0%-62.4%)

ELSE IF *Insulin* 14.5_to_121.0 **THEN** probability of diabetes: 11.7% (6.1%-18.9%)

ELSE IF *BMI* 28.5_to_inf **AND** *Age* 28.5_to_inf **THEN** probability of diabetes: 46.9% (37.2%-56.8%)

ELSE probability of diabetes: 14.5% (7.3%-23.6%)

Fig. 4.28 Rules generated from Bayesian rule lists for the diabetes dataset

Table 4.14 Explainable properties of BRL

Properties	Values
Local or global	Global and local
Linear or non-linear	Non-linear
Monotonic or non-monotonic	Non-monotonic
Feature interactions captured	Yes
Model complexity	Medium to high

Observations:

- The numerical data is discretized into categories using Fayyad and Irani MDL method. Each category, for example, (*BMI* -inf_to_26.45-inf_to_26.45) can be read as the range for the condition $-\inf \leq BMI \leq 26.45$.
- Figure 4.28 The **THEN** part captures the probability of the label (diabetes) and the ranges.
- Interesting to see how high glucose (*glucose* $\geq$ 154.5) and the combination of higher *BMI* (*BMI* $\geq$ 26.45) and being older *Age* (*Age* $\geq$ 28.5) results in higher probability of diabetes in an interpretable manner for any subject matter expert.

Explainable properties of BRL are shown in Table 4.14.

4.6   Scoring System

Many applications such as healthcare, criminal justice systems, finance, to name some, need scoring systems where features are assigned weights that are integers and contribute to the overall scores or help classify. The typical approach is to have a panel of experts set weights and use the data for validating the results. Another standard methodology is to perform feature selection and logistic regression with sparsity constraints, do manual scaling and rounding of the coefficients.

4.6.1   Supersparse Linear Integer Models

One of the ways to address this is to have the objective defined with constraints on accuracy, sparsity, and the coefficients being integers [UR15].

$$\underbrace{\arg\min_{\lambda \in L} \frac{1}{N} \sum_{i=1}^{N} \mathbb{1}[y_i \lambda^T \mathbf{x}_i \leq 0]}_{\text{Accuracy}} + \underbrace{C_0 ||\lambda||_0}_{\text{Sparsity constraint}} + \underbrace{C_1 ||\lambda||_1}_{\text{Co-prime Coefficients constraints}}$$

(4.32)

where $\lambda \in L \implies \forall j \; \lambda_j = \{-10, -9, \ldots, 0, \ldots, 9, 10\}$. The second L_1 term does not introduce sparsity but for small values of ϵ makes the integer coefficients co-prime without impacting either the accuracy or sparsity. These equations are then given to linear programming solvers who find the solutions. Risk-Calibrated Supersparse Linear Integer Models (Risk-SLIM) go even further and not only generate scores with sparse integer based coefficients like SLIM but also make the loss calibrated [UR19].

$$\min_{\lambda \in L} \underbrace{\log(1 + e^{-y_i \mathbf{x}_i \lambda})}_{\text{Logistic Loss}} + \underbrace{C ||\lambda||_0}_{\text{Model Size}}$$

(4.33)

where $\lambda \in L \implies \forall j \; \lambda_j = \{-10, -9, \ldots, 0, \ldots, 9, 10\}$. This objective makes the problem even harder as it is mixed-integer non-linear optimization problem and there is no known solver that solves it reasonably well in a given time. As shown in Fig. 4.29, even with convex objective like the logistic loss, a coefficient through cutting plane approximation can result in a non-integer value, e.g., $\lambda = 4.8$. To address the non-integer coefficients caused by the cutting planes, the objective function is further sub-divided into two problems with bounds between the value, i.e., ($\lambda \leq 4, \lambda \geq 5$) for solution leading into a new algorithm—Lattice Cutting Plane Algorithm (LCPA). To further improve the performance of LCPA, various techniques such as Polishing, SequentialRounding, and Discrete coordinate descent (DCD) are applied to reach an optimal solution referred to as 1-opt solution. Figure 4.30 illustrates the iterative process of SequentialRounding, where one non-integer coefficient is rounded either to an upper bound or a lower bound at a time to see the performance impact (loss reducing) and then leading to the discrete coordinate descent. The integer coefficients from SequentialRounding further go through an iterative process to vary one coefficient at a time to reduce the loss the most. The process stops when there is no further change to any coefficients and leading to an optimal solution known as 1-opt solution.

Interpretation of SLIM: SLIM and Risk-SLIM both output features and their coefficients as points between the given range, for, e.g., $(-5, 5)$. The positive coefficient and the value have the impact of increasing the logistic score and inverse for the negative coefficients.

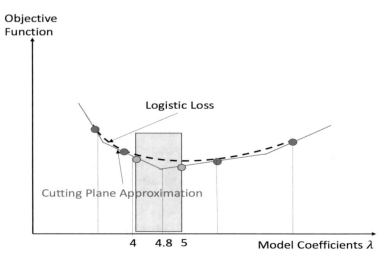

Fig. 4.29 Cutting plane algorithms with bounds for non-convex stalling issues

Model Coefficients λ: λ_0 λ_1 λ_2 λ_3 λ_4 λ_5

	λ_0	λ_1	λ_2	λ_3	λ_4	λ_5	
	1	2	4.8	3	9.7	5	*SequentialRounding*
	1	2	4	3	9.7	5	
	1	2	4	3	10	5	Discrete Coordinate Descent (DCD)
	1	2	4	2	10	5	
	1	1	3	2	10	5	1-opt solution

Fig. 4.30 Iterative process of sequential rounding and discrete coordinate descent to reach the 1-opt solution

```
+-------------------------------------------------+--------------------+--------------+
| Pr(Y = +1) = 1.0/(1.0 + exp(-(-2 + score)))     |                    |              |
| ================================================| ================== | ============ |
| 'Glucose=\'(159.5-179.2]'                       |           2 points | + ....... |
| 'DiabetesPedigreeFunction=\'(0.79-1.01-inf)\''  |           1 points | + ....... |
| 'Age=\'(-inf-27)\''                             |          -1 points | + ....... |
| 'Glucose=\'(79.6-99.5]\''                       |          -1 points | + ....... |
| 'BMI=\'(20.13-26.84]\''                         |          -2 points | + ....... |
| ================================================| ================== | ========== |
| ADD POINTS FROM ROWS 1 to 5                     |              SCORE | = ...... |
+-------------------------------------------------+--------------------+--------------+
```

Fig. 4.31 Scoring the features of diabetes dataset using Risk-SLIM technique

Table 4.15 Explainable properties of Risk-SLIM

Properties	Values
Local or global	Global
Linear or non-linear	Linear
Monotonic or non-monotonic	Monotonic
Feature interactions captured	Yes
Model complexity	Medium to high

Observations:

- Figure 4.31 shows the features *Glucose, DiabetesPedigreeFunction, Age* and *BMI* are the top contributors to the diabetes scoring.
- The numerical data is discretized into categories using Fayyad and Irani MDL method and then binarized to have value (0/1).
- High *Glucose* (159.5–179.2) and having a pedigree with diabetes *Pedigree-DiabetesFunction* increase the score by 2 and 1 points, respectively.
- Similarly, being young (*Age* < 27), lower *Glucose* (79.6–99.5), and BMI in the healthy range (20.13–26.84)) result in lowering the risk score by 1, 2 and 2 points, respectively.

Explainable properties of Risk-SLIM are shown in Table 4.15.

References

[Ade+18] J. Adebayo, et al., Sanity checks for saliency maps. Adv. Neural Inf. Proc. Syst. **31**, 9505–9515 (2018)

[ANS20] G. Aglin, S. Nijssen, P. Schaus, Learning optimal decision trees using caching branch-and-bound search, in *Proceedings of the AAAI Conference on Artificial Intelligence*, vol. 34, No. 04 (2020), pp. 3146–3153

[AS08] A.I.R.L. Azevedo, M.F. Santos, KDD, SEMMA and CRISP-DM: a parallel overview, in *IADS-DM* (2008)

[Bas+18] S. Basu, et al., Iterative random forests to discover predictive and stable high-order interactions. Proc. Nat. Acad. Sci. **115**(8), 1943–1948 (2018)

[BB96] K.P. Bennett, J.A. Blue, Optimal decision trees. Rensselaer Polytechnic Institute Math. Rep. **214**, 24 (1996)

[BD17] D. Bertsimas, J. Dunn, Optimal classification trees. Mach. Learn. **106**(7), 1039–1082 (2017)

[Bre01] L. Breiman, Mach. Learn. **45**(1), 5–32 (2001). ISSN: 0885-6125

[CGM02] H.A. Chipman, E.I. George, R.E. McCulloch, Bayesian treed models. Mach. Learn. **48**(1–3), 299–320 (2002)

[CGM+10] H.A. Chipman, E.I. George, R.E. McCulloch, et al., BART: Bayesian additive regression trees. Ann. Appl. Statist. **4**(1), 266–298 (2010)

[CB91] P. Clark, R. Boswell, Rule induction with CN2: Some recent improvements, in *European Working Session on Learning* (Springer, Berlin, 1991), pp. 151–163

[Coh95b] W.W. Cohen, Fast effective rule induction, in *Machine Learning Proceedings 1995* (Elsevier, Amsterdam, 1995), pp. 115–123

[DMS98] D.G.T. Denison, B.K. Mallick, A.F.M. Smith, A Bayesian cart algorithm. Biometrika **85**(2), 363–377 (1998)

[FW98] E. Frank, I.H. Witten, Generating accurate rule sets without global optimization, in *Proceedings of the Fifteenth International Conference on Machine Learning*, (1998), pp. 144–151

[FS97] Y. Freund, R.E. Schapire, A decision-theoretic generalization of on-line learning and an application to boosting. J. Comput. Syst. Sci. **55**(1), 119–139 (1997)

[FSA99] Y. Freund, R. Schapire, N. Abe, A short introduction to boosting. J. Jpn. Soc. Artif. Intell. **14**(771–780), 1612 (1999)

[Fri00] J.H. Friedman, Greedy function approximation: a gradient boosting machine. Ann. Statist. **29**, 1189–1232 (2000)

[FP+08] J.H. Friedman, B.E. Popescu, et al., Predictive learning via rule ensembles. Ann. Appl. Statist. **2**(3), 916–954 (2008)

[Für99] J. Fürnkranz, Separate-and-conquer rule learning. Artif. Intell. Rev. **13**(1), 3–54 (1999)

[Gar+17] F. Gardin, et al., *skope-rules* (2017). https://github.com/scikit-learn-contrib/skope-rules

[HT90b] T.J. Hastie, R.J. Tibshirani, *Generalized Additive Models*, vol. 43 (CRC Press, Boca Raton, 1990)

[HHZ06] T. Hothorn, K. Hornik, A. Zeileis, Unbiased recursive partitioning: a conditional inference framework. J. Comput. Graph. Statist. **15**(3), 651–674 (2006)

[HS07] W.-C. Hsiao, Y.-S. Shih, Splitting variable selection for multivariate regression trees. Statist. Probab. Lett. **77**(3), 265–271 (2007)

[HRS19] X. Hu, C. Rudin, M. Seltzer, Optimal sparse decision trees, in *Advances in Neural Information Processing Systems* (2019), pp. 7267–7275

[KRS14] B. Kim, C. Rudin, J. Shah, The Bayesian case model: A generative approach for case-based reasoning and prototype classification, in *Proceedings of the 27th International Conference on Neural Information Processing Systems - Volume 2*. NIPS'14 (MIT Press, Cambridge, 2014), pp. 1952–1960

[KL01] H. Kim, W.-Y. Loh, Classification trees with unbiased multiway splits. J. Amer. Statist. Assoc. **96**(454), 589–604 (2001)

[LS04] D.R. Larsen, P.L. Speckman, Multivariate regression trees for analysis of abundance data. Biometrics **60**(2), 543–549 (2004)

[Lar+18] N. Larus-Stone, et al., Systems optimizations for learning certifiably optimal rule lists, in *SysML Conference* (2018)

[Let+15] B. Letham, et al., Interpretable classifiers using rules and Bayesian analysis: building a better stroke prediction model. Ann. Appl. Statist. **9**(3), 1350–1371 (2015)

[LS97] W.-Y. Loh, Y.-S. Shih, Split selection methods for classification trees. Statist. Sinica **7**, 815–840 (1997)

[LV88] W.-Y. Loh, N. Vanichsetakul, Tree-structured classification via generalized discriminant analysis. J. Amer. Statist. Assoc. **83**(403), 715–725 (1988)

[Mic83b] R.S. Michalski, A theory and methodology of inductive learning, in *Machine Learning* (Elsevier, Amsterdam, 1983), pp. 83–134

[Nor+19] H. Nori, et al., InterpretML: a unified framework for machine learning interpretability (2019). Preprint arXiv:1909.09223

[Nor+15] M. Norouzi, et al., Efficient non-greedy optimization of decision trees, in *Advances in Neural Information Processing Systems* (2015), pp. 1729–1737

[Rud19c] R.J. Quinlan, Learning with continuous classes, in *5th Australian Joint Conference on Artificial Intelligence* (World Scientific, Singapore, 1992), pp. 343–348

[Rud19a] C. Rudin, Do simpler models exist and how can we find them? in *KDD* (2019), pp. 1–2

[Rud19b] C. Rudin, Stop explaining black box machine learning models for high stakes decisions and use interpretable models instead. Nat. Mach. Intell. **1**(5), 206–215 (2019)

[Seg92] M. Robert Segal, Tree-structured methods for longitudinal data. J. Amer. Statist. Assoc. **87**(418), 407–418 (1992)

[SS12] R.J. Sela, J.S. Simonoff, RE-EM trees: a data mining approach for longitudinal and clustered data. Mach. Learn. **86**(2), 169–207 (2012)

[UR15] B. Ustun, C. Rudin, Supersparse linear integer models for optimized medical scoring systems. Mach. Learn. **102**(3), 349–391 (2015)

[UR19] B. Ustun, C. Rudin, Learning optimized risk scores. J. Mach. Learn. Res. **20**(150), 1–75 (2019). http://jmlr.org/papers/v20/18-615.html

[Wan+15] T. Wang, et al., Or's of And's for Interpretable Classification, with Application to Context-Aware Recommender Systems (2015). arXiv: 1504.07614 [cs.LG]

[Was+00] L. Wasserman, et al., Bayesian model selection and model aver-aging. J. Math. Psychol. **44**(1), 92–107 (2000)

[YL99] Y. Yu, D. Lambert, Fitting trees to functional data, with an application to time-of-day patterns. J. Computat. Graph. Statist. **8**(4), 749–762 (1999)

Chapter 5
Post-Hoc Interpretability and Explanations

Post-hoc techniques represent a vast collection of methods created to specifically address the black-box problem, where we do not have access to the internal feature representations or model structure. There are considerable advantages to using post-hoc methods. They can work for a wide variety of model algorithms. They allow for different representations to be used for internal modeling and explanation. They can also provide different types of explanations for the same model. However, there is a trade-off between the fidelity and comprehensibility of explanations.

We divide post-hoc methods into three separate categories. Visual explanations use figures to express and interpret relationships between model features and predictions in a human readable fashion. Feature importance methods allow us to quantify these relationships in a more precise manner, whether they be local explanations on specific instances or global explanations on the overall model. Example-based explanations seek to explain models using the prediction of individual instances of the model. They can provide insight on nuances in the data distribution that could be missed by other methods. Applying a collection of these methods together can provide meaningful, intuitive explanations for even the most complex models.

5.1 Tools and Libraries

Table 5.1 provides details of all the libraries used for various models in the chapter.

5.2 Visual Explanation

Visual explanation encompasses a set of methods that allow us to visually inspect the relationships between inputs and outputs or with other input features. This allows

Table 5.1 Models and implementations

Model/algorithm	Library
Partial dependence	Scikit-Learn
Individual conditional dependence	Scikit-Learn
Ceteris paribus	pyCeterisParibus
Accumulated local effects	ALEPython
Breakdown	dalex
iBreakdown	dalex
Feature interaction	Scikit-Learn
Permutation feature importance	Scikit-Learn
Ablations LOCO	lofo-importance
Kernel SHAP	shap 0.39.0
Anchors	alibi
Global surrogate	Scikit-Learn
LIME	lime-0.2.0.1
Contrastive explanations	alibi
kNN	Scikit-Learn
Trust scores	alibi
Counterfactuals	alibi
Prototypes/criticisms	MMD-critic
Influential instances	Scikit-Learn

us to better understand their influence and contribution to the predicted response. Visual explanations have the advantage in that they are relatively easy to interpret, especially for models with a small set of features. They can provide global or local explanations, depending on the method. However, they can become unwieldy with a larger feature set and may not properly capture or visualize correlations between features, resulting in erroneous attributions. They may also extrapolate poorly for regions not well covered by the training data. We describe and discuss six visual explanation methods below.

5.2.1 Partial Dependence Plots

The partial dependence (PD) plot is commonly used as a visual explanation of feature effects. It is a global method and can explain the overall model behavior by showing the relationships between input features and output. PD plots visualize the partial dependence function which measures the effect of a feature by marginalizing over other features. For instance, given a model with two features, $f(x_1, x_2)$, the partial dependence function of feature x_1 is given by averaging over the marginal distribution of feature x_2:

$$\text{PD}(x_1) = \mathbb{E}_{x_2}[f(x_1, x_2)] = \int p(x_2) f(x_1, x_2) \, dx_2 \qquad (5.1)$$

where $p(x_2)$ represents the marginal distribution of feature x_2. In practice, we can replace the expectation with an average over n data samples where we replace the first feature by x_1:

$$\text{PD}(x_1) = \frac{1}{n} \sum_{j=1}^{n} f\left(x_1, x_{2,j}\right) \tag{5.2}$$

For a model with a set of d features $x_1, x_2, \ldots, x_d$, the PD function for x_1 would be computed as

$$\text{PD}(x_1) = \frac{1}{n} \sum_{j=1}^{n} f\left(x_1, x_{2,j}, \ldots, x_{d,j}\right) \tag{5.3}$$

As seen above, marginalizing over the other features allows the PD function to capture interactions of other features. It shows on average how feature x_1 influences the model output. The PD plot shows the PD function for all possible values of feature x_1.

We can extend PD plots to show the partial dependence of two or more features at the same time. Let the features to be plotted be defined as the subset $C \subseteq \{x_1, x_2, \ldots, x_d\}$, and let $\overline{C}$ be the other features of the model such that $C \cup \overline{C} = \{x_1, x_2, \ldots, x_d\}$. The PD function for subset C is written as

$$\text{PD}(C) = \mathbb{E}_{\overline{C}}\left[f(C, \overline{C})\right] \tag{5.4}$$

$$= \frac{1}{n} \sum_{j=1}^{n} f(C, \overline{C_j}) \tag{5.5}$$

where $\overline{C}_j$ are the other feature values for data sample j. When there are two features of interest, they can be plotted in two dimensions as a two-dimensional PD plot. For more than two features, visualizing a multidimensional plot is more difficult.

PD plots have many advantages. They provide a clear and understandable explanation of how the output of a model varies with changes to a particular feature. They are easy and quick to compute. PD plots also illuminate the causal relationship between the PD feature and the model output, as they are based on the actual model output for given values of the PD feature. They can also display the partial dependence of two features concurrently as a two-dimensional plot.

Interpretation of Partial Dependence Plots: PD plots are a simple way to visualize the individual effects of features on the prediction outcome. They average over features based on the marginal distribution. They can be used

(continued)

to visualize the partial dependence and interaction of two features on model prediction. They implicitly assume independence between the PD feature/s and the other model features.

Unfortunately, PD plots inherently assume that the PD feature is uncorrelated with the other features. Any co-dependence between features x_1 and x_2 is unaccounted for and PD plots will be biased and depict unrealistic out-of-sample instances when the features are correlated. This problem worsens with more correlated features. PD plots also do not provide information about the distribution of features, which can be detrimental when the distribution is imbalanced. Due to

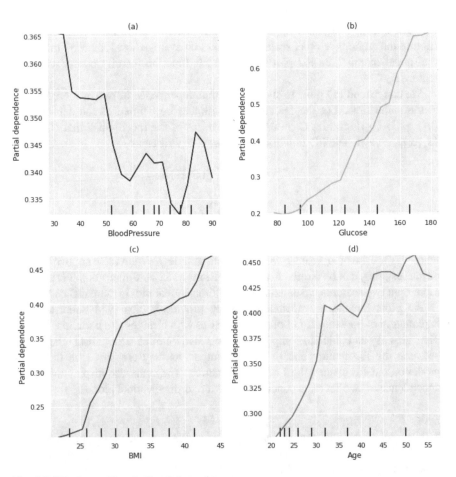

Fig. 5.1 PD plots on Pima Indian diabetes dataset

the averaging, marginal effects are smoothed and any heterogeneous effects in the data (where subgroups of data samples behave differently) can be overlooked.

Explainable properties of Partial Dependence Plots are shown in Table 5.2.

Observations:

- Figure 5.1 shows that *Glucose* and *BMI* have a direct effect on model outcome, with less effect by *Age* and little effect by *BloodPressure*. Higher *Glucose* and *BMI* indicate higher probability of diabetes.
- Figure 5.2 indicates a positive interaction effect of *BMI* and *Glucose* on model prediction for diabetes.
- Figure 5.3 shows the positive interaction effect of *Insulin* and *Glucose* on model prediction as a surface plot.

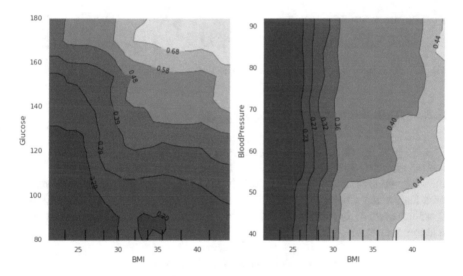

Fig. 5.2 2D PD plot on Pima Indian diabetes dataset

Table 5.2 Explainable properties of partial dependence plots

Properties	Values
Local or global	Global
Linear or non-linear	Linear
Monotonic or non-monotonic	Monotonic
Feature interactions captured	No
Model complexity	Low

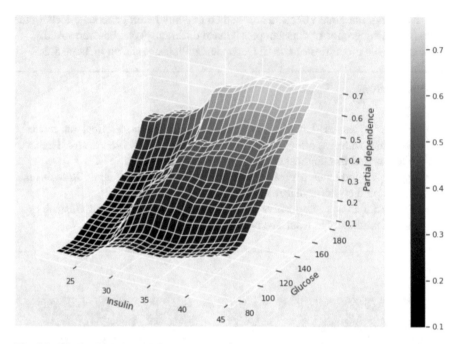

Fig. 5.3 3D PD plot on Pima Indian diabetes dataset

5.2.2   *Individual Conditional Expectation Plots*

The PD plot shows the average overall effect, but there are times that we want to see the effect on each data sample. The Individual Conditional Expectation (ICE) plot is analogous to the PD plot in that it visualizes the relationship of the output on a feature but plots a separate line for each instance. In similar fashion, the ICE plot for feature x_1 is computed by replacing the first feature while keeping all other features the same:

$$\text{ICE}(x_1) = f\left(x_1, x_{2,j}, \ldots, x_{d,j}\right)|_{j=1,\ldots,n} \tag{5.6}$$

Whereas partial dependence plots average out and obscure possible hetero-geneous effects in the data samples, ICE plots can clearly show when distinct subgroups behave differently. Furthermore, ICE plots can show when correlations exist between the plotted feature and the remaining features. If such interactions exist, the ICE plot will be significantly more insightful.

One downside of ICE plots is that for large data samples, they appear crowded. This is exacerbated when the lines start at different output prediction values. Centering is one way to reduce the noise, by translating the lines for the data samples such that they start at the same value and plotting only the differences in model output beyond this point.

> Interpretation of Individual Conditional Expectation Plots: ICE plots show the partial dependence of model prediction on a feature for every data instance in the dataset as an individual line. The mean of these lines is the partial dependence function. ICE plots can reveal heterogeneous behavior of subsets of data. They also assume independence between the feature plotted and the remaining model features.

While more intuitive than PD plots to understand, ICE plots can only show a single feature at a time. A two-dimensional ICE plot would be extremely difficult to comprehend unless the data sample size were very small. Furthermore, even though ICE plots can be helpful in discerning possible correlations between the plotted feature and other features, they still suffer from the same bias as PD plots in which points may depict unrealistic out-of-sample instances.

Explainable properties of Individual Conditional Expectation Plots are shown in Table 5.3.

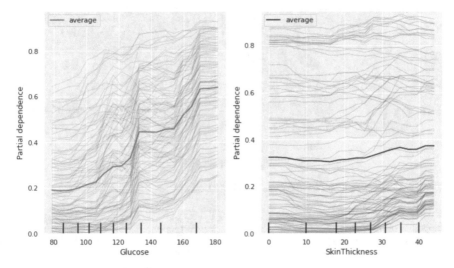

Fig. 5.4 ICE plot on Pima Indian diabetes dataset

Observations:

- Figure 5.4a shows the ICE plot for *Glucose*. The PDP line is the average of all lines.
- Heterogeneous effects can be observed in *SkinThickness* in Fig. 5.4b, where the individual ICE lines are distributed more densely below vs. above the PDP average.
- By centering the ICE lines in Fig. 5.5, heterogeneous effects are more easily observable.

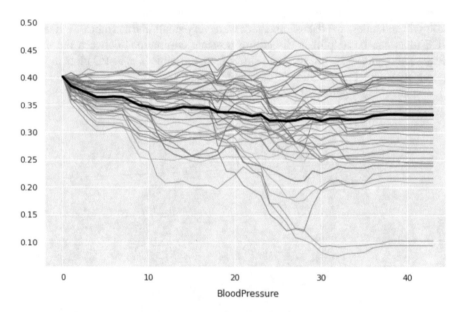

Fig. 5.5 Centered ICE plot on Pima Indian diabetes dataset

Table 5.3 Explainable properties of individual conditional expectation plots

Properties	Values
Local or global	Local
Linear or non-linear	Linear
Monotonic or non-monotonic	Monotonic
Feature interactions captured	No
Model complexity	Low

5.2.3   Ceteris Paribus Plots

The Ceteris Paribus (CP) plot is named after the Latin phrase "ceteris paribus" meaning "all other things being equal." Like the PD and ICE plots, it is useful for explaining the influence of the features of a model on its output. Specifically, the CP plot shows the relationship between the change in model prediction and the changes in feature values for a single data instance. This is in contrast to PD and ICE plots, which show the change in individual features across all data instances.

Given a model $f(x)$ with d features, the one-dimensional CP plot for feature i at point of interest x^* is given by

$$CP_i(z)|_{x^*} = f\left(x_1^*, x_2^*, \ldots, x_i^* + z, \ldots, x_d^*\right) \qquad (5.7)$$

Thus, the CP plot describes the influence on the model prediction by the value of the i-th feature at the specific instance x^*, with all other features held constant. In practice, $CP_i(z)$ is plotted over the entire range of observed values of the i-th feature, and it is easier to visualize a set of features on a single CP plot.

CP plots are particularly useful for evaluating model sensitivity and model comparison. Comparing CP plots for two or more instances can reveal the sensitivity of the model. By overlaying plots for two models, the differences in feature influence between the models become apparent.

Interpretation of Ceteris Paribus Plots: CP plots are useful for explaining the feature influence on model prediction for a specific data instance. They assume independence between the feature plotted and the remaining model features.

Unfortunately, CP plots make the same underlying assumption as PD plots in that little to no correlation exists between model features. For dependent features, a change in one will occur with a change in another, which is not captured by CP plots.

Explainable properties of Ceteris Paribus Plots are shown in Table 5.4.

Observations:

- Figure 5.6 reveals that *Glucose* and *BMI* have the largest influence on diabetes prediction, whereas *SkinThickness* has the least influence.
- The CP plots also indicate that *BloodPressure* and *Insulin* levels have an inverse relationship with model outcome, such that higher levels of either lead to lower diabetes prediction.

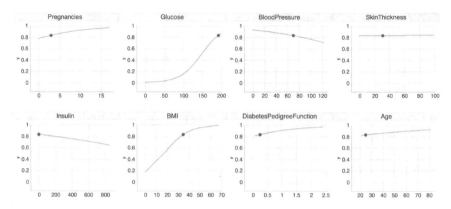

Fig. 5.6 Ceteris Paribus plot on Pima Indian diabetes dataset

Table 5.4 Explainable
properties of Ceteris Paribus
plots

Properties	Values
Local or global	Local
Linear or non-linear	Linear
Monotonic or non-monotonic	Monotonic
Feature interactions captured	No
Model complexity	Low

5.2.4   Accumulated Local Effects Plots

Accumulated local effects (ALE) plots are similar to PD plots in that they visualize
the average relationship between features and the output of a machine learning
model. However, PD plots are biased when correlations between features are
significant. ALE plots are an alternative that can account for potential correlations
and feature interactions. Given a model f with two features, the ALE function for
feature x_1 is given by

$$\text{ALE}(x_1) = \int_{\min(x_1)}^{x_1} \mathbb{E}\left[f(x_1, x_2) | x_1 = z\right] dz - C \tag{5.8}$$

$$= \int_{\min(x_1)}^{x_1} \int p(x_2|z) \frac{\partial f(z, x_2)}{\partial z} dx_2 dz - C \tag{5.9}$$

where C is a constant chosen to vertically center the plot such that the mean effect
is zero:

$$C = \mathbb{E}\left[\text{ALE}(x_1)\right] \tag{5.10}$$

As seen in this equation, ALE averages over the features using the conditional
distribution $p(x_2|x_1)$ rather of the marginal distribution $p(x_2)$ as in PD, which

avoids the extrapolation of data to unrealistic combinations of feature values as in PD plots. Instead of averaging model predictions, ALE averages over the change in model predictions $\partial f(x_1, x_2)/\partial x_1$, which represents the local effect of x_1 on $f(x_1, x_2)$. This effectively blocks and offsets possible correlations that might exist between feature x_1 and other features. These local effects are averaged across all possible values of feature x_2. The integral accumulates the local effects over values of z up to the value of x_1, hence the name "accumulated local effects."

Method	Distribution	Average over
PD	Marginal	Model predictions
ALE	Conditional	Changes in model predictions

In practice, we split feature x_1 into a set of K segments that end at values $x_{1,1}, x_{1,2}, \ldots, x_{1,K}$ and average over n data samples while using finite differences to calculate the change in model prediction:

$$\text{ALE}(x_1) = \sum_{k=1}^{K} \frac{1}{n(k)} \sum_{i=1}^{n(k)} \left[f(x_{1,k}, x_{2,i}) - f(x_{1,k-1}, x_{2,i}) \right] - C \tag{5.11}$$

where $n(k)$ is the number of data samples that lie within the k-th segment of x_1 and $x_{2,i}$ is the value of the second feature for the i-th data sample that lies within the k-th segment. The constant C is given by

$$C = \frac{1}{n} \sum_{i=1}^{n} \text{ALE}(x_1) \tag{5.12}$$

Thus, given a feature value, the relative effect of changing the ALE feature on model output can be directly read from the ALE plot. The ALE value at each point is interpreted as the difference from the average output prediction.

ALE plots have significant computational advantages over PD plots. For a set of N features, K segments, and n data samples, ALE plots require $2^N \times n$ model predictions, whereas PD plots require $K^N \times n$ predictions. Since the computation of ALE plots does not depend on K, the number of computations scales linearly as the number of samples n increases.

The equivalent of ICE plots to PD plots does not exist for ALE plots. This is because each interval used to calculate ALE plots uses only samples that fall within the interval. Additionally, ALE plots average the difference in effects between samples, and it would be impossible to plot a curve for each sample.

Although ALE plots can account for correlated features, it can still be difficult to interpret them when the features are strongly correlated. In such cases, it is better to either plot only the dominant feature or plot the joint changes in the correlated features together. Furthermore, as the number of intervals K increases and fewer data points lie within each interval, ALE plots can become noisy in appearance. Careful consideration should be taken in the selection of K.

> Interpretation of Accumulated Local Effects Plots: ALE is useful for visualizing the relationship between one or two features and model output. They take into account feature interactions since they average using conditional distributions and are much faster to compute in comparison to PD plots.

Analogous to two-dimensional PD plots, 2D ALE plots can be used to show the interaction of two features at once. For a model $f(x_1, x_2, \ldots, x_d)$ on d features, the 2D ALE function is given by

$$\text{ALE}(x_1, x_2) = \int_{\min(x_1)}^{x_1} \int_{\min(x_2)}^{x_2} \mathbb{E}\left[f(x_1, x_2, \ldots, x_d)|x_1 = z, x_2 = q\right] dz dq - C$$
(5.13)

where the constant C is chosen to doubly center the ALE function such that the mean effect of features x_1 and x_2 is zero:

$$C = \frac{1}{n} \sum_{i=1}^{n} \text{ALE}(x_1, x_2)$$
(5.14)

It is important to note that 2D ALE plots explicitly show the second-order effects of the two features on model prediction. If no such interaction exists, the 2D ALE plot will be close to zero. This is very different from the PD plot as 1D and 2D PD plots always show the total effect.

Explainable properties of Accumulated Local Effects Plots are shown in Table 5.5.

> **Observations:**
>
> - Figure 5.7 shows the direct relationship of *Glucose* levels and inverse relationship of *Insulin* levels on diabetes prediction, which states that higher *Glucose* or lower *Insulin* leads to higher probability of diabetes.
> - We can observe the second-order interaction effects of *Glucose* and *Insulin* together in Fig. 5.8, where concurrently lower levels of both have less interaction effect than concurrently higher levels.

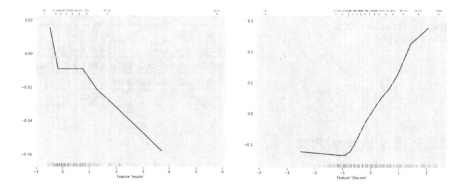

Fig. 5.7   ALE plot on Pima Indian diabetes dataset

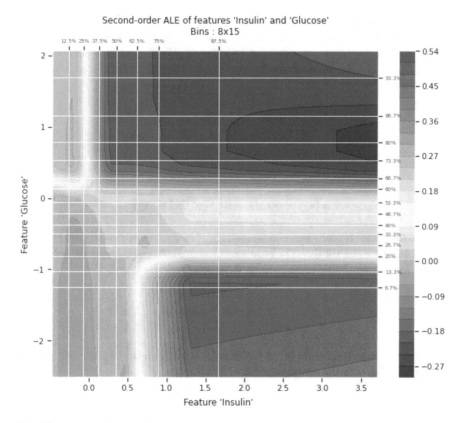

Fig. 5.8   2D ALE plot on Pima Indian diabetes dataset

Table 5.5 Explainable
properties of accumulated
local effects plots

Properties	Values
Local or global	Global
Linear or non-linear	Non-linear
Monotonic or non-monotonic	Non-monotonic
Feature interactions captured	Yes
Model complexity	Low

5.2.5   Breakdown Plots

Breakdown plots visualize the influence of model features on the model prediction
by decomposing the prediction into individual attributions of each feature for a
specific data instance. It is a local method that explains which features have the
strongest influence on a particular prediction.

Breakdown plots measure the change in model output for a change in each feature
while holding the values of other features constant. Given a model with d features,
the approach quantifies the attribution $v_i(x)$ of feature i at instance $x = x^*$ relative
to the mean model prediction $f(\overline{x})$:

$$f(x^*) = f(\overline{x}) + \sum_{i=1}^{d} v_i(x^*) \tag{5.15}$$

with

$$v_i(x^*) = f(x^*) - \mathbb{E}\left[f(x^*)|x_i^* = X_i\right] \tag{5.16}$$

and the expectation taken over the random variable X_i. We call $v_i(x^*)$ the variable-
importance measure for feature i evaluated at sample x^*. For linear additive models
of the form

$$f(x) = \alpha + \sum_{i=1}^{d} \beta_i x_i \tag{5.17}$$

the expectation of a function is the function of an expectation. We can then replace
the expectation with the sample mean:

$$v_i(x^*) = f(x^*) - f\left(\frac{1}{n}\sum_{i=1}^{n} x_i\right) \tag{5.18}$$

$$= \beta_i\left(x_i^* - \overline{x}_i\right) \tag{5.19}$$

Interpretation of Breakdown Plots: breakdown plots explain a model prediction on a particular instance by quantifying feature attribution. Individual feature attributions are calculated by holding the values of other features constant. Changing order in which feature attributions are calculated can significantly change explanations if feature interactions exist.

In general, breakdown plots provide order-specific explanations of feature attributions. For linear additive models, the order in which breakdown plots select and evaluate variable-importance measures does not matter. However, for nonlinear models, in which feature interactions exist, differences in order can lead to significantly different explanations for the same instance.

Explainable properties of Breakdown Plots are shown in Table 5.6.

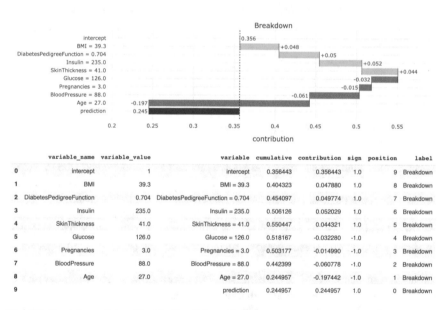

	variable_name	variable_value	variable	cumulative	contribution	sign	position	label
0	intercept	1	intercept	0.356443	0.356443	1.0	9	Breakdown
1	BMI	39.3	BMI = 39.3	0.404323	0.047880	1.0	8	Breakdown
2	DiabetesPedigreeFunction	0.704	DiabetesPedigreeFunction = 0.704	0.454097	0.049774	1.0	7	Breakdown
3	Insulin	235.0	Insulin = 235.0	0.506126	0.052029	1.0	6	Breakdown
4	SkinThickness	41.0	SkinThickness = 41.0	0.550447	0.044321	1.0	5	Breakdown
5	Glucose	126.0	Glucose = 126.0	0.518167	-0.032280	-1.0	4	Breakdown
6	Pregnancies	3.0	Pregnancies = 3.0	0.503177	-0.014990	-1.0	3	Breakdown
7	BloodPressure	88.0	BloodPressure = 88.0	0.442399	-0.060778	-1.0	2	Breakdown
8	Age	27.0	Age = 27.0	0.244957	-0.197442	-1.0	1	Breakdown
9			prediction	0.244957	0.244957	1.0	0	Breakdown

Fig. 5.9 Breakdown plot on Pima Indian diabetes dataset

Table 5.6 Explainable
properties of breakdown plots

Properties	Values
Local or global	Local
Linear or non-linear	Linear
Monotonic or non-monotonic	Monotonic
Feature interactions captured	No
Model complexity	Low

Observations:

- Fig. 5.9 quantifies the feature attribution for the prediction of a particular instance. We see that 3 features contribute positively and 4 negatively toward diabetes prediction.
- Age had the most significant effect in reducing diabetes prediction.
- Taken together, the features lead to a 0.245 prediction probability for diabetes.

5.2.6 Interaction Breakdown Plots

When feature interactions exist, breakdown plots are not suitable and lead to widely varying, biased explanations. Interaction Breakdown plots (iBreakdown) capture feature interactions while generating explanations for feature contributions. To do so, iBreakdown bootstraps data samples to a baseline model with fixed parameters and uses breakdown plots to generate a set of breakdown explanations. The variation in these explanations serves a proxy to model level uncertainty. It also permutes feature order in the breakdown plots, where the variation of feature contribution values is a measure of the feature explanation uncertainty. A greater variability of contribution implies interaction with other features.

For a model $f(x)$, let the individual contribution Δ_i for feature i at instance x^* be given by

$$\Delta_i = \mathbb{E}\left[f(x)|x_i = x_i^*\right] - \mathbb{E}\left[f(x)\right] \tag{5.20}$$

For each pair of features (x_i, x_j), their joint contribution is given by

$$\Delta_{ij} = \mathbb{E}\left[f(x)|x_i = x_i^*, x_j = x_j^*\right] - \mathbb{E}\left[f(x)\right] \tag{5.21}$$

The interaction contribution Δ_{ij}^I for each pair of features (x_i, x_j) is given by

$$\Delta_{ij}^I = \Delta_{ij} - \Delta_i - \Delta_j \tag{5.22}$$

This interaction contribution will be zero if no interaction exists between the pair of features. By using the individual contributions Δ_i and interactive contributions Δ_{ij} to rank and sequentially order the features, breakdown variable-importance scores can be applied to compute feature attribution.

Explainable properties of iBreakdown Plots are shown in Table 5.7.

Interpretation of iBreakdown Plots: iBreakdown is an improvement upon breakdown plots to handle the effects of feature interactions. It provides ranking and ordering of both individual features and interaction effects for breakdown plots.

Observations:

- Fig. 5.10 shows feature attributions for the same instance as in Fig. 5.9, except that feature interactions are also incorporated.
- The figure indicates that *BMI*, *DiabetesPedigreeFunction*, *Insulin*, and *SkinThickness* contributed positively to diabetes prediction.
- It can be seen that the interaction effect of *BloodPressure* and *Glucose* had the strongest negative contribution to diabetes prediction.

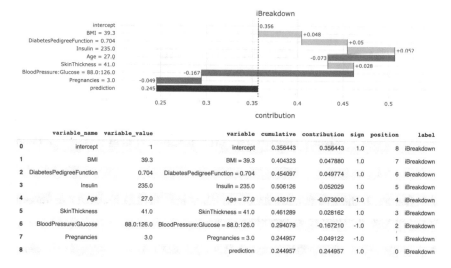

	variable_name	variable_value	variable	cumulative	contribution	sign	position	label
0	intercept	1	intercept	0.356443	0.356443	1.0	8	iBreakdown
1	BMI	39.3	BMI = 39.3	0.404323	0.047880	1.0	7	iBreakdown
2	DiabetesPedigreeFunction	0.704	DiabetesPedigreeFunction = 0.704	0.454097	0.049774	1.0	6	iBreakdown
3	Insulin	235.0	Insulin = 235.0	0.506126	0.052029	1.0	5	iBreakdown
4	Age	27.0	Age = 27.0	0.433127	-0.073000	-1.0	4	iBreakdown
5	SkinThickness	41.0	SkinThickness = 41.0	0.461289	0.028162	1.0	3	iBreakdown
6	BloodPressure:Glucose	88.0:126.0	BloodPressure:Glucose = 88.0:126.0	0.294079	-0.167210	-1.0	2	iBreakdown
7	Pregnancies	3.0	Pregnancies = 3.0	0.244957	-0.049122	-1.0	1	iBreakdown
8			prediction	0.244957	0.244957	1.0	0	iBreakdown

Fig. 5.10 iBreakdown plot on Pima Indian diabetes dataset

Table 5.7 Explainable
properties of iBreakdown
plots

Properties	Values
Local or global	Local
Linear or non-linear	Non-linear
Monotonic or non-monotonic	Non-monotonic
Feature interactions captured	Yes
Model complexity	Medium

5.3   Feature Importance

Analogous to visual explanation methods, feature importance methods seek to
quantify the contribution of features on model prediction. They do so while
considering additional factors such as the type, robustness, comprehensibility, or
quality of explanations. Explanations may explain the global behavior of a model,
or they may be local and explain the prediction of individual instances. Each method
has its advantages and trade-offs, and some are model-specific or optimized to
particular model methods. We explore a variety of these methods below.

5.3.1   Feature Interaction

When the features of a model are not independent, they will interact with each other
to influence the model prediction. We can subtract the total effect of all features
$f(x_*)$ from the sum of the effects of each individual feature $f(x_i)$ to find the feature
interaction effect $f_*(x)$:

$$f_*(x) = \sum_{i=1}^{d} f(x_i) - f(x_*) \qquad (5.23)$$

Let the partial dependence function of two features (x_1, x_2) be given by

$$\mathrm{PD}(x_1, x_2) = \frac{1}{n} \sum_{k=1}^{n} f(x_1, x_2, x_{3,k}, \ldots, x_{d,k}) \qquad (5.24)$$

where the summation is over the data and $x_{d,k}$ is the k-th data instance value of the
d-th feature. If the two features have no interaction, this PD function decomposes
to the sum of individual PD functions:

$$\mathrm{PD}(x_1, x_2) = \mathrm{PD}(x_1) + \mathrm{PD}(x_2) \qquad (5.25)$$

Hence, another way to measure the interaction effect $\mathrm{PD}_*(x_1, x_2)$ between these
features is given by

$$PD_*(x_1, x_2) = PD(x_1, x_2) - PD(x_1) - PD(x_2) \qquad (5.26)$$

The two-way H statistic proposed by [FP08] is a measure of feature interaction based on the variance of the interaction effect between two features x_j and x_k:

$$H_{jk}^2 = \frac{\sum_{i=1}^{n} \left[PD(x_{j,i}, x_{k,i}) - PD(x_{j,i}) - PD(x_{k,i}) \right]^2}{\sum_{i=1}^{n} PD^2(x_{j,i}, x_{k,i})} \qquad (5.27)$$

where the summation is once again over the data and $PD(x_{j,i})$ is the partial dependence function of the j-th feature at the i-th data instance value. It can be interpreted as the share of variance explained by the interaction effect between two features. A variation of the H statistic is based on the variance of the interaction effect between the j-th feature and all of the remaining features $\bar{j}$:

$$H_{j}^2 = \frac{\sum_{i=1}^{n} \left[f(x_i) - PD(x_{j,i}) - PD(x_{\bar{j},i}) \right]^2}{\sum_{i=1}^{n} f^2(x_i)} \qquad (5.28)$$

This is termed the interaction H statistic.

We can calculate the interaction H statistic for each feature in a model to measure their interaction effect with the other model features and then explore the relationships using the two-way H statistic.

> Interpretation of Feature Interaction: H statistic plots can allow us to measure the interaction effect between a feature and the model output or between two features. It can be interpreted as the contribution to variance by the interaction effect. It is computationally expensive and cannot explain the nature of the interaction.

The H statistic has meaningful interpretation as the contribution to variance by the interaction effect. Because it ranges between zero and one depending on the strength of interaction between features x_j and x_k or of the j-th feature and the remaining features, it is useful for comparison across varying feature types and models. The trade-off is that the H statistic is very computationally expensive, as the PD function for every pair of features must be computed over n data samples. Also, the H statistic can measure the strength but not the type of interaction. They cannot be used to relate the change in model prediction from changes in the input features.

Explainable properties of Feature Interaction are shown in Table 5.8.

Observations:

- Figure 5.11 shows the ranked interaction H statistic between the features and diabetes prediction.
- *Glucose* is seen to have the highest interaction, whereas *BloodPressure* has the lowest interaction.
- The two-way H statistic between pairs of features is shown in Fig. 5.12.
- It is not unexpected that *Pregnancies* and *Age* have the strongest relationship.
- What is somewhat surprising is that *BloodPressure* and *SkinThickness* have such a strong a relationship.

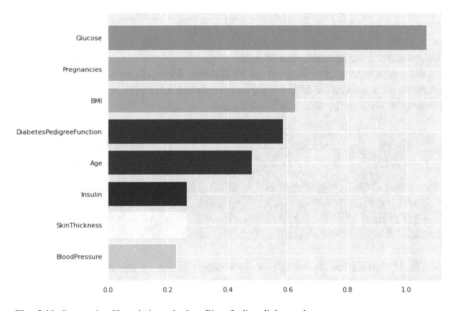

Fig. 5.11 Interaction H statistic ranked on Pima Indian diabetes dataset

Table 5.8 Explainable properties of feature interaction

Properties	Values
Local or global	Global
Linear or non-linear	Non-linear
Monotonic or non-monotonic	Non-monotonic
Feature interactions captured	Yes
Model complexity	High

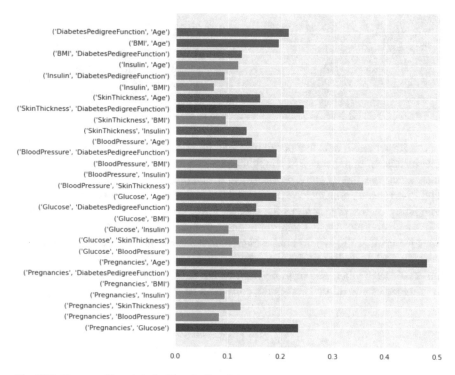

Fig. 5.12 Two-way H statistic for Pima Indian diabetes dataset

5.3.2 Permutation Feature Importance

Permutation feature importance (PFI) was proposed as a global explanation method. It is based on the notion that the importance of a feature is related to the rise in model prediction error when the values of the feature are shuffled. The greater the error, the stronger the relationship between the feature and the model prediction.

Consider a model f with d features. Given two features (x_i, x_j), if we permute the values of feature x_i and observe the model error rises relative to when we permute the values of feature x_j, we conclude that x_i is more important. For linear additive models, this implies the magnitude of the weight of feature x_i is larger than for x_j. For non-linear models, the interpretation of feature permutation is less clear, since interaction effects between features are included and cannot be isolated when applying PFI.

The interpretation of PFI is easy to understand and quick to compute since it does not require models to be retrained. PFI can be calculated as an absolute difference in model error (e.g., difference in mean squared error) or as a ratio of model errors. Using a ratio allows PFI to be applied across a variety of models. However, PFI requires the true predictions to be known in order to calculate the model error.

Interpretation of Permutation Feature Importance: PFI is based on the notion that shuffling the values of unimportant features does not increase model error. It is computationally efficient since no models are retrained. In general, PFI should be calculated on test set data. PFI is biased when features are strongly correlated.

It is important to note that permuting features can often lead to unrealistic, invalid data instances. In general, PFI should be computed on test data rather than training data, since the model may be overfit and computing PFI on training data may lead to erroneous conclusions about which features are most important. However, computing on test data ignores the training instances, resulting in greater variance. PFI can be combined with cross-validation to reduce variance, but computing on the folds means that PFI would not be computed using all the data on the final model. Because PFI includes all feature interaction effects, feature importance may be underestimated when two or more features are strongly correlated. This is because importance can be diluted across the correlated features.

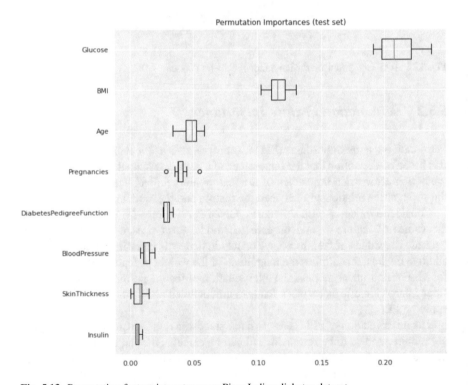

Fig. 5.13 Permutation feature importance on Pima Indian diabetes dataset

Table 5.9 Explainable properties of permutation feature importance

Properties	Values
Local or global	Global
Linear or non-Linear	Non-linear
Monotonic or non-monotonic	Non-monotonic
Feature interactions captured	Yes
Model complexity	Low

Explainable properties of Permutation Feature Importance are shown in Table 5.9.

Observations:

- Figure 5.13 shows that *Glucose*, *BMI*, and *Age* have the highest importance, whereas *BloodPressure*, *SkinThickness*, and *Insulin* have the lowest.
- Because *Glucose* and *Insulin* may be strongly negatively correlated, the importances of both may be biased and *Insulin* may be underestimated whereas *Glucose* overestimated.

5.3.3 Ablations: Leave-One-Covariate-Out

Ablation is a group of methods that sequentially remove features and evaluate the change in model prediction error to determine feature relevance. One of the simplest methods is called Leave-One-Covariate-Out (LOCO), also known as Leave-One-Feature-Out (LOFO). As its name implies, the idea is to sequentially drop each feature, retrain the model, and compare the subsequent model error with respect to the baseline model that includes all features. This comparison can be in the form of the difference in prediction error (e.g., mean squared error) or the ratio of model errors.

For a model f with d features and n data samples, LOCO as a ratio for the j-th feature is given by

$$LOCO_j = \frac{\sum_{i=1}^{n} \left| y_i - f(x_{1,i}, x_{2,i}, \ldots, x_{j-1,i}, x_{j+1,i}, \ldots, x_{d,i}) \right|^2}{\sum_{i=1}^{n} \left| y_i - f(x_{1,i}, x_{2,i}, \ldots, x_{d,i}) \right|^2} \quad (5.29)$$

where y_i is the true value of the i-th data instance and $x_{j,i}$ is the j-th feature for the i-th data instance. Like permutation feature importance, LOCO needs access to the true outputs for computation. Because LOCO requires the model with d features to be retrained d times, computational costs can be huge when the feature space is large.

Interpretation of Leave-One-Covariate-Out: LOCO has simple and easily understandable interpretation, but it is computationally expensive since it requires a model to be trained as many times as there are features. Feature interaction effects are also ignored.

The results of LOCO are simple to understand and interpret. LOCO is applicable across a variety of feature and model types. However, as LOCO only drops a single feature at a time, it ignores feature interaction effects when two or more features are correlated. If interaction effects are significant, LOCO will return incorrect or biased results.

Explainable properties of Leave-One-Covariate-Out are shown in Table Table 5.10.

Observations:

- Figure 5.14 shows that *Glucose* is the single most significant feature for diabetes prediction, whereas *Pregnancies* is the least important.
- LOCO is fairly certain that *Glucose* is the most significant, while it is quite uncertain that *Age* is the third most significant feature.

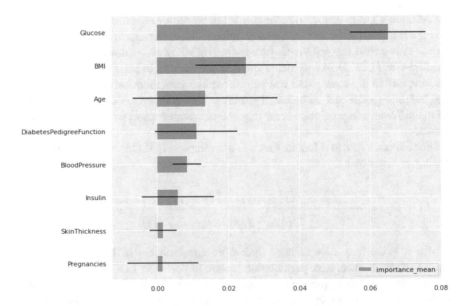

Fig. 5.14 Leave-One-Covariate-Out ranked on Pima Indian diabetes dataset

Table 5.10 Explainable properties of Leave-One-Covariate-Out

Properties	Values
Local or global	Global
Linear or non-linear	Non-linear
Monotonic or non-monotonic	Non-monotonic
Feature interactions captured	No
Model complexity	Medium

5.3.4 Shapley Values

Shapley values have game theoretic foundations and are another useful method for measuring feature relevance. They are based on the notion of distributing feature importance between participating features by applying cooperative game theory. Each feature is a player in a game whose goal is to distribute payouts fairly among players based on their contribution to the total payout. A coalition is a group of players cooperating together in the game. The Shapley value is defined as the marginal contribution of each feature averaged across the set of all possible coalitions of features. It was shown by its namesake Shapley [Sha53] in 1953 to be efficient, symmetric, dummy, and additive, which implies a fair distribution.

Given a model f with d features $X = \{x_1, x_2, \ldots, x_d\}$, the Shapley value ϕ_j for the j-th feature is given by

$$\phi_j(f) = \sum_{S \in \mathcal{P}(X \setminus \{x_j\})} \frac{|S|!(d - |S| - 1)!}{d!} \left[f\left(S \cup \{x_j\}\right) - f(S) \right] \tag{5.30}$$

where the summation is over the power set of all possible feature coalitions $S \in \mathcal{P}(X \setminus \{x_j\})$ that do not contain the j-th feature, and $|S|$ is the number of elements (features) in coalition S. Here, $f\left(S \cup \{x_j\}\right)$ and $f(S)$ are models trained with feature x_j present and absent, respectively. The Shapley value can be written as

$$\phi_j(f) = \frac{1}{d} \sum_{S \in \mathcal{P}(X \setminus \{x_j\})} \frac{f\left(S \cup \{x_j\}\right) - f(S)}{\binom{d-1}{|S|}} \tag{5.31}$$

It can be seen from this equation that the Shapley value is the average marginal contribution of adding feature x_j to all possible coalitions excluding x_j. The model prediction can thus be expressed as a summation of the Shapley values and the mean model prediction:

$$f(x) = \sum_{j=1}^{d} \phi_j(f) + \mathbb{E}\left[f(x)\right] \tag{5.32}$$

In this fashion, Shapley values explain the difference between the model prediction and the global average prediction. Shapley values possess several useful properties, including local accuracy (where the explanation model matches the original model for the instance to be explained), missingness (where the explanation will be zero for an irrelevant feature), and consistency (where the explanation will not decrease if a feature's contribution does not decrease). These properties imply that there is a unique additive feature attribution method based on Shapley values.

Unfortunately, Shapley values do not adequately handle feature interactions. Furthermore, with a large feature space, Shapley values have high computation cost since the number of coalitions increases exponentially.

> Interpretation of Shapley Values: use Shapley values to explain feature contributions for specific instances. Note that they do not handle feature interactions well and are difficult to compute exactly.

5.3.5 SHAP

One of the difficulties in calculating Shapley values is the exponential computational cost. For practical considerations, SHAP (Shapley Additive Explanations) was proposed by [LL17b] to explain a particular data instance as a unified measure of feature importance. It provides a unique, additive feature importance explanation for individual data instances.

Recall that the marginal contribution $C\left(x_j|S\right)$ of a feature x_j to a coalition S is given by

$$C\left(x_j|S\right) = f\left(S \cup \{x_j\}\right) - f\left(S\right) \qquad (5.33)$$

SHAP computes Shapley values of a conditional expectation function of the model over a dataset:

$$f(X) = \mathbb{E}\left[f(X)|S\right] \qquad (5.34)$$

where the model features are given by $X = S \cup (X \setminus S)$. As it is expensive to retrain a large set of models on subsets of features, SHAP applies the observed marginal expectation to calculate model predictions:

$$f(S) = \mathbb{E}\left[f(X)|S = S^*\right] \qquad (5.35)$$

where S^* are the feature values of the data instance to be explained. The expectation is taken over a background dataset, which requires a sufficient number of samples of both S and $X \setminus S$ to achieve accurate results.

As computation of the conditional expectation is challenging, several methods have been proposed to approximate the calculation of SHAP values. We discuss one of these in the next section.

5.3.6 KernelSHAP

KernelSHAP is an algorithm that computes approximate Shapley values in the local region around a reference data instance X^* by

1. sampling over all possible coalitions and
2. applying a linear additive model approximation to f.

Because of a linear model assumption, the features are uncorrelated and the SHAP conditional expectation can be approximated by

$$f(S) = \mathbb{E}\left[f\left(S^* \cup (X \setminus S)\right)\right] \tag{5.36}$$

This is significantly computationally cheaper to calculate and implies we can compute model predictions by fixing the feature values of S to those of the sample to be explained X^* and sampling the values of the features not in S from the background dataset.

Computing KernelSHAP is equivalent to fitting a linear model by minimizing the objective function $\mathcal{J}$:

$$\mathcal{J} = \min_{\phi_i,\dots,\phi_M} \left\{ \sum_{S \in \mathcal{P}(X \setminus \{x_j\})} \left[f(S) - \sum_{x_j \in S} \phi_j \right]^2 \pi(S) \right\} \tag{5.37}$$

where the kernel $\pi(S)$ is given by

$$\pi(S) = \frac{d-1}{\binom{d}{|S|} |S| (d - |S|)} \tag{5.38}$$

Since f is assumed to be linear, and $\mathcal{J}$ is a square loss function, least squares regression can be used. It can be seen that the kernel represents a weighting function that assigns higher weight to the smallest and largest coalitions. This is because they tend to be most informative in determining a feature importance, and a coalition with one feature is as informative as a coalition with all but one feature.

In practice, we calculate Shapley values using KernelSHAP and average across the entire dataset:

$$\Phi_j = \sum_{i=1}^{n} \left| \phi_j^i \right| \tag{5.39}$$

Interpretation of SHAP: SHAP provides an efficient way to calculate Shapley values to help explain the feature contributions for a specific data instance. KernelSHAP is one model-agnostic implementation of SHAP. KernelSHAP computes approximate Shapley values and does not account for feature interactions.

Shapley values predicted by KernelSHAP are approximate. Furthermore, due to its linear model assumption, KernelSHAP does not account for feature interactions. However, we can apply KernelSHAP concepts to calculate the Shapley interaction effect, given by

$$\phi_{i,j} = \sum_{S \in \mathcal{P}(X \setminus \{x_i, x_j\})} \frac{|S|!(d - |S| - 2)!}{2(d-1)!} \delta_{ij}(S) \tag{5.40}$$

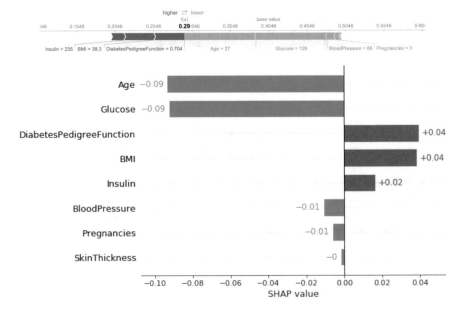

Fig. 5.15 SHAP on Pima Indian diabetes dataset

where $i \neq j$ and the interaction contribution is given by

$$\delta_{ij}(S) = f(S \cup \{x_i, x_j\}) - f(S \cup \{x_i\}) - f(S \cup \{x_j\}) + f(S) \qquad (5.41)$$

This relation isolates the interaction effect by removing the individual effects for features x_i and x_j.

Explainable properties of KernelSHAP are shown in Table 5.11.

Observations:

- Figure 5.15 indicates that *Age* and *Glucose* have the greatest negative influence on the instance prediction, while *DiabetesPedigreeFunction* and *BMI* have the greatest positive influence.
- Figure 5.16 shows the distribution of the Shapley values over all instances, from which we see that *Glucose* and *BMI* have the largest average impact on the prediction of diabetes.

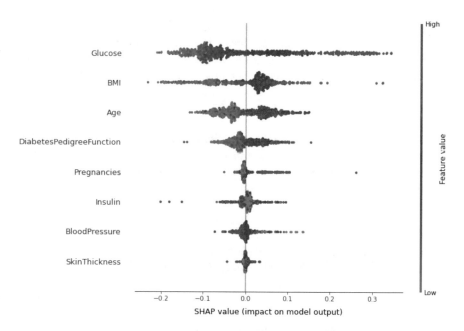

Fig. 5.16 Shapley values predicted by KernelSHAP for all data instances in Pima Indian diabetes dataset

Table 5.11 Explainable
properties of KernelSHAP

Properties	Values
Local or global	Local
Linear or non-linear	Linear
Monotonic or non-monotonic	Monotonic
Feature interactions captured	No
Model complexity	Medium

5.3.7 Anchors

Precision is a key factor in interpretability, as humans prefer explanations with high precision and fidelity. Anchors are a high-precision method for generating local explanations by extracting a set of if-then rules. These rules have the property in that they serve as "anchors" for a set of features such that changes in other features will not change the model output. Anchor explanations thus represent local, high-confidence, sufficient conditions for predictions. They are explicitly clear in their explanation since they apply only when all of their rules are met. Anchors do not assume local linearity as in KernelSHAP or LIME and are therefore more precise in handling unseen data instances.

With complex models, explaining individual predictions can make the explanation more comprehensible. Local explanations commonly generate perturbation-based explanations that measure the change in model output for small changes to the data instances. Let D represent a known distribution of perturbations. Given a model f, instance x, and precision threshold τ, an anchor A is given by

$$\mathbb{E}_{D(z|A)}\left[\mathbb{1}_{f(x)=f(z)}\right] \geq \tau, \quad A(x) = 1 \tag{5.42}$$

where z is a sample drawn from the conditional distribution when rule A applies $D(z|A)$. In other words, an anchor is a set of feature predicates on x that achieves a precision $\rho(A) \geq \tau$. In general, anchors are expected with high probability to meet this precision threshold for some arbitrarily small $/delta$:

$$\Pr\left[\rho(A) \geq \tau\right] \geq 1 - \delta \tag{5.43}$$

It may be that several anchors meet this criterion. In this case, preference is given to the anchors with the highest coverage $\mathcal{C}(A)$, defined as

$$\mathcal{C}(A) = \mathbb{E}_{D(z)}\left[A(z)\right] \tag{5.44}$$

If we define A' to be the set of anchors that satisfy (5.43), generation of anchors can be achieved by solving the combinatorial optimization problem:

$$\max_{A \in A'} \mathcal{C}(A) \tag{5.45}$$

Note that the anchor search space is exponential.

In practice, a bottom-up anchor construction approach is taken that starts with an empty rule set and generates a set of candidate rules by iteratively extending an anchor by one additional feature predicate. The anchor is replaced with the rule with the highest estimated precision at each step. Once the precision threshold is met, the anchor is added to the set of anchors for the model. This is a greedy search approach that will identify anchors with the fewest rules and not necessarily the highest coverage, but such anchors are easier to understand and generally have higher coverage. Beam search improves upon the greedy search method by maintaining a candidate set of rules during the iterative search process.

> Interpretation of Anchors: Anchors are local, high-confidence, rules-based explanations for specific instances that explicitly specify their coverage. A set of anchors with high coverage over the dataset can help us understand global model behavior.

Anchors overcome limitations of other local explanation methods that model local behavior using linear approximations. For complex models and data spaces, this linear approximation may be invalid and lead to inaccurate results. Furthermore, the region applicable to a local linear model is unclear, whereas anchors explicitly specify their coverage. However, anchor rules may sometimes be complex and provide low coverage. They also do not provide as much insight beyond identifying instances near decision boundaries. A set of generated anchors may have overlapping coverage and/or omit a portion of the input space. Lastly, anchors require a perturbation distribution that should be realistic, interpretable, and expressive enough to uncover model behavior.

Explainable properties of Anchors are shown in Table 5.12.

Observations:

- Figure 5.17 lists four anchors (two for positive and two for negative diabetes prediction) generated from the dataset.
- The first anchor states that if *Glucose* >139 and *BMI* >32.35 and *Age* >40, diabetes is predicted with 98% percent precision.
- The two negative anchors are based on lower values of *Glucose* and *BMI* and have significant overlap. In general, we prefer to use the second anchor, which has higher coverage (16.9 vs 9.3%).
- The second positive anchor is very specific and incorporates criteria for six features but suffers from very low coverage (0.16%).

Table 5.12 Explainable
properties of anchors

Properties	Values
Local or global	Local
Linear or non-linear	Non-linear
Monotonic or non-monotonic	Non-monotonic
Feature interactions captured	Yes
Model complexity	High

anchor	precision	coverage	prediction
Glucose > 139.00 AND BMI > 32.35 AND Age > 40.00	0.978814	0.065147	1
Glucose <= 99.00 AND BMI <= 27.40	1.000000	0.092834	0
Glucose > 117.00 AND BMI > 36.60 AND Age > 29.00 AND BloodPressure <= 64.00 AND DiabetesPedigreeFunction > 0.37 AND Insulin > 125.00	0.982456	0.001629	1
BMI <= 27.40 AND Glucose <= 117.00	1.000000	0.169381	0

Fig. 5.17 Anchors on Pima Indian diabetes dataset

5.3.8 Global Surrogate

Global surrogate methods attempt to approximate the predictions of a black-box model by using a simpler, more interpretable model in order to explain model behavior. This simpler model is termed the surrogate model and is fitted to emulate the input–output mapping of the black-box model. Explanations of the surrogate model can then serve to explain the global behavior of the black-box model.

Consider a black-box model f, a data instance x, and a surrogate model h. Fitting the global surrogate model is equivalent to the optimization problem:

$$\underset{h}{\arg\min}\ \mathbb{E}_x\left[\mathbb{E}_{f(x)}\left[L(h(x), f(x))\right]|x\right] \tag{5.46}$$

where L is a specified loss function over the joint distribution of the black-box model inputs and outputs. If the loss function is mean squared error, the fidelity of the surrogate model to the black-box model can be measured by R-squared (R^2):

$$R^2 = 1 - \frac{\sum_{i=1}^{n}\left[h(x_i) - f(x_i)\right]^2}{\sum_{i=1}^{n}\left[\overline{h} - f(x_i)\right]^2} \tag{5.47}$$

where the summation is over a dataset of n samples, $h(x_i)$ and $f(x_i)$ are the surrogate model and black-box model predictions on the i-th sample, respectively, and where $\overline{h}$ is the mean surrogate model prediction given by

$$\overline{h} = \frac{1}{n}\sum_{i=1}^{n} h(x_i) \tag{5.48}$$

This R^2 is a measure of how well the black-box model variance is explained by the surrogate model, and a value close to 1 indicates a good fit.

Applying global surrogate methods in practice generally follows the following steps:

1. Select an interpretable model (e.g., linear/logistic regression, decision tree, Naive Bayes, etc.).
2. Create a dataset from the inputs and black-box model outputs.
3. Train the interpretable model on the dataset.
4. Use the interpretable model explanations to understand black-box model behavior.

> Interpretation of Global Surrogate Models: global surrogates aim to explain complex models by fitting simpler, interpretable models to the complex model predictions. The simpler model can then be used to explain the global behavior of the more complex model.

Global surrogate models generally exhibit a complexity vs precision trade-off. While they are easier to compute and interpret, they may lack sufficient power to accurately capture complex models or data distributions. A global surrogate model trained on one data sample may make wildly divergent predictions for another data sample. Furthermore, applying different interpretable models may result in different explanations for the same black-box model. Global surrogate models generally cannot make accurate predictions on individual data instances. Even if the surrogate model predictions perfectly mirror the black-box model, there is no guarantee that the black-box model predictions match the real world. Surrogate models would provide only an illusion of interpretability.

Explainable properties of Global Surrogates are shown in Table 5.14.

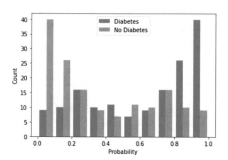 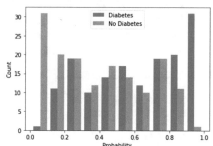

Fig. 5.18 Prediction histogram of logistic regression (global surrogate model) vs random forest model for diabetes classification on Pima dataset (R-squared $= 0.70$)

Table 5.13 Feature weights of global surrogate (logistic regression) model

Feature	Weight
Pregnancies	0.11307387
Glucose	1.25676028
BloodPressure	−0.28624197
SkinThickness	−0.05920553
Insulin	−0.06240799
BMI	0.87315981
DiabetesPedigreeFunction	0.29223447
Age	0.3340456

Table 5.14 Explainable properties of global surrogates

Properties	Values
Local or global	Global
Linear or non-linear	Both
Monotonic or non-monotonic	Both
Feature interactions captured	Yes
Model complexity	Low

Observations:

- Figure 5.18 on the left shows prediction histogram of surrogate model (logistic regression) fitted to the predictions of a black-box model (random forest). The right side shows the prediction histogram of the black-box model.
- A comparison of the histograms indicates that the model predictions are comparable, though the surrogate model is more sure of its prediction with a greater number of predictions toward the ends. The two models exhibited an R-squared of 0.70 and a correlation of 84%.
- We can use the feature weights of the surrogate model to explain black-box model prediction. As seen in Table 5.13, *Glucose* and *BMI* have the great influence on diabetes prediction.

5.3.9 LIME

Local surrogate models aim to explain the predictions of a black-box model on individual data instances rather than on the entire model. They exhibit a property called local fidelity, which means that they are good approximations to how a black-box model behaves in the vicinity of the instance to be predicted. In 2016, Local Interpretable Model-Agnostic Explanations (LIME) was proposed [RSG16a] as a model-agnostic approach to generate interpretable explanations of the individual prediction of any model in the region around the prediction. Specifically, LIME is based on

- using interpretable representations of data instances that are human understandable,
- approximating the black-box model in the neighborhood of a data instance with an interpretable (linear) model,
- perturbing the data instance of interest to generate new samples weighted by their proximity to the instance of interest,
- fitting the interpretable model to the new samples and using it to provide explanations of the black-box model.

Formally, for a black-box model f, data instance of interest x, and an explanation model h, the problem of identifying a local surrogate model is equivalent to optimizing for the objective $\zeta(x)$:

$$\zeta(x) = \arg\min_{h \in \mathscr{H}} \mathscr{L}(f, h, \pi_x(z)) + \Omega(h) \tag{5.49}$$

where π_x is a kernel function that weighs a sample z based on its distance to the instance of interest x, $\mathscr{L}$ is a loss function, and *Omega* is a complexity penalty. The explanation model h is selected from a class $\mathscr{H}$ of interpretable models such as linear, logistic, or decision tree models. The complexity term $\Omega(h)$ is selected to constrain the number of model parameters or tree depth by including a cost for greater complexity. The loss function $\mathscr{L}$ represents a measure of how poorly the explanation model approximates the black-box model in the locality of instance x.

In practice, the loss $\mathscr{L}$ is approximated by applying random perturbations to the instance of interest to generate a set of new samples z. These samples are weighted by the kernel function $\pi_x(z)$ that penalizes the distance between x and z. This kernel weighting allows LIME to be fairly robust to sampling noise.

An important element of LIME is the mapping of data instances to interpretable representations that are easily understandable to humans, such as the use of binary vectors to represent the presence or absence of a feature or a set of features. For example, each element of a binary vector can represent a value, a word, or a patch of pixels, and a value of 0 or 1 would indicate the absence or the presence, respectively, within the data instance to be mapped. Interpretable representations have drawbacks. Information may be lost when transforming the data instances, they may lack sufficient representation power, or they may constrain the interpretable model.

Interpretation of Local Interpretable Model-Agnostic Explanations: LIME provides human-interpretable feature relevance explanations for individual

(continued)

instances. It does so by randomly perturbing instances and training a local surrogate model on these perturbations which is most often a linear model. As a model-agnostic method, LIME is very flexible and applicable across different domains and data types.

LIME is very flexible and generates simple, easy to understand models. However, it may not reveal the complete explanation if the behavior of the black-box model is highly non-linear near the instance of interest. Sampling by random perturbation may not provide sufficient coverage, and it is possible to generate completely different explanations for two data instances that are near each other. For high dimensionality, the notion of distance between any two points becomes less well defined and care must be taken in defining the kernel weighting function within LIME as well as pre-scaling across dimensions. Lastly, recent work by [Sla+20] has shown that LIME, SHAP, and other input perturbation-based explanation methods can make (and be fooled into providing) innocuous explanations for black-box models with hidden biases.

Explainable properties of LIME are shown in Table 5.15.

Observations:

- Figure 5.19 depicts the LIME explanations for one instance with a high probability (79%) prediction of diabetes.
- *Glucose* and *BMI* can be seen to be the most influential features contributing to the positive prediction.
- *BloodPressure* was the sole negative contribution.

Fig. 5.19 LIME applied to classification on Pima dataset

Table 5.15 Explainable
properties of LIME

Properties	Values
Local or global	Local
Linear or non-linear	Both
Monotonic or non-monotonic	Both
Feature interactions captured	Yes
Model complexity	Medium

5.4 Example-Based

Example-based explanations are a set of model-agnostic methods that attempt to explain global or local behavior of a model or its underlying data distribution by selecting particular data instances. Unlike previous methods, example-based methods do not explain models through feature importance or interactions. Instead, they help humans construct compact mental models of the black-box model or of the data distribution.

5.4.1 Contrastive Explanation

The contrastive explanation method (CEM), also known as differential explanations, does not explain a specific prediction but instead aims to explain why a prediction was made in contrast to another. It generates local explanations for the prediction of black-box classifier models on individual instances. CEM is based on the notion of pertinent positives (PPs) and pertinent negatives (PNs). Whereas pertinent positives are the minimal set of relevant features that must be present in order to justify the prediction, pertinent negatives are the minimum set of features that must be absent in order to assert the prediction of the instance of interest. Taken together, PP and PN constitute a complete explanation and evidence in support of the model prediction.

Given a black-box model f, a dataset D, an instance of interest x with d features, its associated prediction $t = f(x)$, and a scoring function $\zeta(x)$ that returns a set of confidence scores over all classes, the CEM method identifies pertinent positives by optimizing over the perturbation function δ^{pos} in the expression:

$$
\delta^{pos} \leftarrow \underset{\delta \in \Delta_{pp}}{\arg \min} \; \alpha \max \left[\underbrace{\underset{i \neq t}{\max} \, [\zeta(x+\delta)]_i - \underbrace{[\zeta(x+\delta)]_t}_{\text{predicted class}}, -\kappa} \right]
$$
$$\underbrace{\qquad\qquad\qquad}_{\text{next likely class}}$$
$$\underbrace{\qquad\qquad\qquad\qquad\qquad\qquad}_{\text{hinge loss function}}$$

(5.50)

$$
+ \beta \underbrace{||x + \delta - b||_1}_{\text{L1 term}} + \underbrace{||x + \delta - b||_2^2}_{\text{L2 term}} - \underbrace{\gamma p(x+\delta)}_{\text{distribution term}}
$$
$$\underbrace{\qquad\qquad\qquad\qquad\qquad}_{\text{elastic net regularizer}}$$

where α, β, and γ are non-negative parameters used to regularize the loss function, L1 term, and distribution term, respectively. The hinge loss function includes a parameter κ, which controls the gap between the predicted class and the next most likely class based on the scoring function. The distribution term is estimated from the data and penalizes for points outside the feasible data space. The perturbation δ^{pos} is constrained to the space

$$
\Delta_{pp} : \; |x + \delta + b| \preceq |x - b| \tag{5.51}
$$

where the $\preceq$ implies element-wise inequality and $b = [b_1, b_2, \ldots, b_d]$ are the base values associated with each feature in x.

CEM identifies pertinent negatives by optimizing over the perturbation function δ_{neg} in the expression:

$$
\delta^{neg} \leftarrow \underset{\delta \in \Delta_{pn}}{\arg \min} \; \alpha \max \left[\underbrace{\underset{i \neq t}{\max} \, [\zeta(x+\delta)]_t - \underbrace{[\zeta(x+\delta)]_i}_{\text{predicted class}}, -\kappa} \right]
$$
$$\underbrace{\qquad\qquad\qquad}_{\text{next likely class}}$$
$$\underbrace{\qquad\qquad\qquad\qquad\qquad\qquad}_{\text{hinge loss function}}$$

(5.52)

$$
+ \beta \underbrace{||\delta||_1}_{\text{L1 term}} + \underbrace{||\delta||_2^2}_{\text{L2 term}} - \underbrace{\gamma p(x+\delta)}_{\text{distribution term}}
$$
$$\underbrace{\qquad\qquad\qquad}_{\text{elastic net regularizer}}$$

where the perturbation δ^{neg} is constrained to the space

$$
\Delta_{pn} : \; |x + \delta + b| \succ |x - b| \tag{5.53}
$$

This is equivalent to find the minimum set of features in $\delta \in \delta_{pn}$ such that their absence in the instance of interest x will lead to a different prediction (the closest different class).

> Interpretation of Contrastive Explanations: aligned with human reasoning,
> CEM is based on sets of pertinent positive and pertinent negative features
> that together constitute a local explanation of a model prediction. CEM is
> applicable to high-dimensional sparse data and captures feature interactions.

In practice, CEM provides easily interpretable explanations that are suitable and
useful for high-dimensional sparse data, especially since the sparsity implies that
there are many feature values that may contain no information for a model prediction. These explanations tend to be persuasive as they are intuitive and analogous
to human decision-making processes. CEM also handles feature interactions, but it
can be sensitive to noise in data and require significant computation costs.

Explainable properties of Contrastive Explanations are shown in Table 5.16.

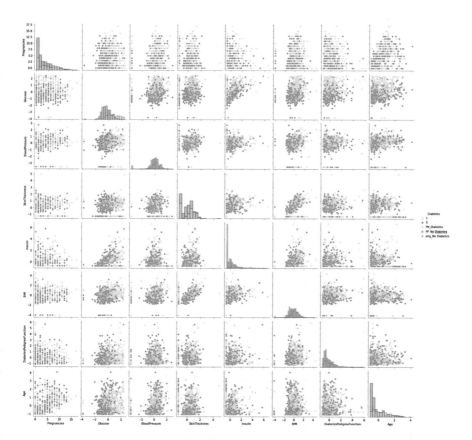

Fig. 5.20 CEM applied to the Pima Diabetes dataset

Table 5.16 Explainable
properties of contrastive
explanations

Properties	Values
Local or global	Local
Linear or non-linear	Both
Monotonic or non-monotonic	Both
Feature interactions captured	Yes
Model complexity	Medium

Observations:

- Figure 5.20 shows pair plots of the model features with the explained instance highlighted in orange, a PP instance in dark green, and a PN instance in light green.
- The PP instance typically lies close to the center of mass of the instances with the same prediction class as the explained instance.
- The PN instance typically lies away from the center of mass and sometimes as an outlier.

5.4.2   kNN

The k-nearest neighbor (kNN) model can be used as an explanatory model to provide local explanations for black-box classifier predictions by assigning the most common class of the k data samples in closest proximity to the data instance of interest. For regression, kNN can be used by assigning the mean output of the k nearest data samples. We are able to draw conclusions as to why the black-box model made its prediction by examining the nearest neighbors. Common measures of proximity used in kNN include cosine, Euclidean, and Minkowski, Manhattan, Mahalanobis, and chi-squared distances.

Because kNN makes no assumptions about the data or black-box model, it can handle non-linear datasets and complex interactions. To do so, however, sufficient data is required, and data sparsity can lead to an insufficient number of instances that are near enough to make accurate predictions. Thus, the choice of k is a parameter that is dependent on the model and dataset. As the kNN algorithm must iterate through all data instances, it suffers from a heavy computational and memory cost, especially for large high-dimensional datasets. Dimension-reduction techniques like PCA are often used beforehand to reduce the computational burden.

Explainable properties of kNN are shown in Table 5.17.

Interpretation of k-Nearest Neighbor Explanations: as a local explanation
method, kNN maps an instance to its k-nearest neighbors and allows us to
draw conclusions based on the predictions of these neighbors. kNN can handle
complex data and models but is computationally expensive and requires
sufficient data to be useful. For high-dimensional data, PCA preprocessing
is almost a necessity.

Observations:

- Figure 5.21 shows the kNN prediction boundary for two features: *Glucose*
  and *BloodPressure*, for levels of $k \in$ 5, 10, 15, 20 relative to the dataset.
- For higher k, outliers have less effect on model prediction.
- kNN can capture complex decision boundaries, though higher k values
  provide a smoothing effect.

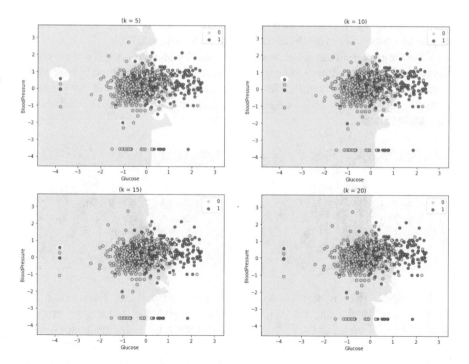

Fig. 5.21 kNN applied Pima Diabetes dataset

Table 5.17 Explainable
properties of kNN

Properties	Values
Local or global	Local
Linear or non-linear	Both
Monotonic or non-monotonic	Both
Feature interactions captured	Yes
Model complexity	High

5.4.3   Trust Scores

The trust score method (TSM) aims to measure the confidence of a black-box classifier model in making a local prediction. It is based on using a modified nearest neighbor classifier to measure the trust score T of an instance of interest x. This score is defined as the ratio between the distance from x to the nearest class that is distinct from the predicted class and the distance from x to the predicted class. Higher trust scores correspond to higher confidence and greater trust in the prediction. While TSM can provide a measure of our trust in a prediction, it does not provide an explanation of the prediction.

In practice, TSM is used in conjunction with other explanation techniques to provide local and global explanations. Data instances with high trust scores have high confidence predictions and likely lie far from the decision boundary. Low trust scores imply ambiguity in predictions, and combining TSM with contrastive explanations or anchors can be more effective in this region.

TSM requires building separate k-d trees for each prediction class trained on the dataset. At inference time, TSM measures the distance of an instance to each of the trees and calculates the ratio of distances from the best next class to the predicted class to generate trust scores. Feature scaling is important to standardize and remove data scale effects.

> Interpretation Trust Scores: TSM provides a measure of the confidence of a model in making a prediction for an instance of interest. This score is a ratio of the distance between the instance to the predicted class and the next nearest class (or other class for binary prediction). TSM is computationally expensive for high-dimensional data, where preprocessing with dimensionality reduction methods like PCA may be required.

Like kNN, TSM is sensitive to noise in data and outliers. These issues can be mitigated by filtering out instance outliers during k-d training or outlying instances at inference time. TSM also suffers from large computational costs on large, high-dimensional datasets, and commonly datasets can be preprocessed with dimensionality reduction techniques to improve computational tractability.

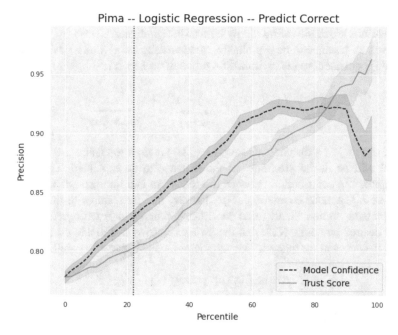

Fig. 5.22 Trust scores applied to Pima Diabetes dataset

Observations:

- Figure 5.22 shows both the trust score and the logistic prediction probability (model confidence) for a Logistic Regression classifier model predicted over a test set.
- We can see from this plot that the logistic regression model tends to overestimate prediction probabilities until 85th percentile, above which the model underestimates the prediction probability.
- Using the trust score allows us to conclude that the model may be more accurate for highly probable predictions.

5.4.4 Counterfactuals

Counterfactuals are defined as thinking about what did not happen but could have happened. It turns out that humans are very adept at applying counterfactual thinking (e.g., "what could be done differently to achieve a different result"). Based on this concept, the counterfactual explanations (CEs) method was proposed. CF generates local explanations by detailing the changes to an input instance that would cause the prediction of a model f to change. Due to their counterfactual nature, CF explanations are intuitive and highly human-interpretable.

Given an instance to be explained x, a counterfactual x', and a desired outcome y', CF is equivalent to the optimization problem on $L(x'|x)$:

$$\min_{x'} L(x'|x) = \underbrace{\left(f(x') - y'\right)^2}_{\text{loss term}} + \lambda \underbrace{d(x, x')}_{\text{distance term}} \qquad (5.54)$$

where $d(x, x')$ is a distance measure (e.g., L1) chosen to ensure that the counterfactual will be in the proximity of the instance to be explained. The parameter λ balances the contribution between the loss term and the distance term. A large value of λ generates counterfactuals with very similar features to the instance to be explained, while a small value generates counterfactuals with predictions close to the desired outcome. Note that the instances are synthetic and do not represent actual instances in the dataset. The parameter λ can be learned as well:

$$\min_{x'} \max_{\lambda} L(x'|x) \text{ with } |f(x) - y'| \le \epsilon \qquad (5.55)$$

where ϵ is a tolerance parameter.

> Interpretation of Counterfactual Explanations: CF methods apply the "what if" principle to generate local explanations. They create nearby synthetic instances with the smallest changes in features that change the model prediction. CF explanations are easy to understand and interpret, though sometimes multiple conflicting explanations may arise.

A limitation of CF is that it can generate many counterfactual explanations all at once. These explanations may be very different or conflict with one another, known as the "Rashomon" effect [FRD19]. When this happens, it would be difficult to select the optimal explanation without additional criteria.

Explainable properties of Counterfactuals are shown in Table 5.18.

Observations:

- Figure 5.23 plots an instance of interest and a generated counterfactual along two features: *Glucose* and *BloodPressure*.
- Note that the counterfactual is a synthetic instance and does not map to an actual instance in a dataset.
- The counterfactual is very near to the explained instance along the *Glucose* axis, while further away along the *BloodPressure* axis.

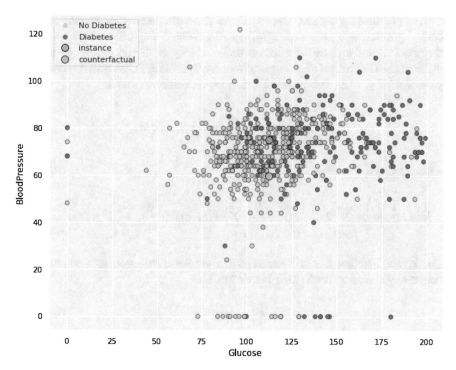

Fig. 5.23 Counterfactuals on the Pima Diabetes dataset

Table 5.18 Explainable
properties of counterfactuals

Properties	Values
Local or global	Local
Linear or non-linear	Both
Monotonic or non-monotonic	Both
Feature interactions captured	Yes
Model complexity	Medium

5.4.5 Prototypes/Criticisms

Prototypes/Criticisms (PC) is a method that seeks to identify instances in the data that are either very representative or very non-representative of the entire dataset, respectively. Studies have shown that the use of prototypes is universally effective for human learning. However, example-based learning can lead to over-generalization or biased predictions. The use of anti-prototypical examples, called criticisms, can help us build better mental models of the data and predictions.

As a practical implementation of PC, the Mean Maximum Discrepancy (MMD)-critic algorithm was proposed. MMD is a measure of the difference between two distributions and is used to measure the similarity of the prototypes to the data distribution. Given a set of n samples from the dataset D, PC identifies an m^* set of prototypes $S \subseteq D$ as a normalized discrete optimization problem:

$$\max_{S \in 2^n, |S| < m^*} \frac{2}{n|S|} \sum_{i,j=1}^{n} k(x_i, x_j) - MMD^2 \qquad (5.56)$$

where MMD is given by

$$MMD^2 = \frac{1}{n^2} \sum_{i,j=1}^{n} k(x_i, x_j) - \frac{2}{nm} \sum_{i,j=1}^{n,m} k(x_i, z_j) + \frac{1}{m^2} \sum_{i,j=1}^{m} k(z_i, z_j) \qquad (5.57)$$

The kernel function $k(x, x')$ measures the similarity between two points and is commonly chosen to be the radial basis function:

$$k(x, x') = \exp^{-\lambda ||x^- x'||^2} \qquad (5.58)$$

A set of c^* criticisms C is identified by the optimization:

$$\min_{C \subseteq [n], |C| < c^*} L(x) + R(x) \qquad (5.59)$$

where $L(x)$ is called the witness function:

$$L(x) = \frac{1}{n} \sum_{i=1}^{n} k(x, x_i) - \frac{1}{m} \sum_{j=1}^{m} k(x, z_j) \qquad (5.60)$$

and $R(x)$ is an optional regularization term. The witness function measures the magnitude of dissimilarity between a set of prototypes and the dataset, and criticisms are the set of points with the largest witness function values.

The prototypes and criticisms method is flexible with inherent benefits in helping us to better understand the distribution of data. It is applicable to all types of data and can be used to generate both local and global explanations for a black-box model. The processes of identifying prototypes and criticisms are independent of each other and can be calculated in parallel.

Interpretation of Prototypes and Criticisms: the PC method identifies two sets of instances in the data: a set of most and a set of least representative instances within the entire dataset. A comparison of these two sets helps explain global model behavior. The MMD-critic algorithm can be used to identify these sets, though it is unclear how to determine the optimal number of prototypes and criticisms in a dataset.

It is unclear how to optimally determine the optimal number of prototypes and criticisms. In general, prototypes will be located in areas with high data density, whereas criticisms are located in areas of low data density. With high-dimensional sparse data, the number of prototypes may need to be high relative to low-dimensional dense data. The number of criticisms is directly related to the number of outliers existing in the data as well.

Explainable properties of Prototypes/Criticisms are shown in Table 5.19.

Observations:

- Figure 5.24 depicts a sample of the prototypes and criticisms generated from the Fashion MNIST dataset using the MMD-critic algorithm.
- The left-side image shows the prototypes associated with the category "t-shirt." They are all recognizable as prototypical instances of t-shirts.
- The right-side image shows the criticisms associated with the category "t-shirt." These rather unusual designs are least representative of the dataset.

Fig. 5.24 Prototypes and criticisms on Fashion MNIST dataset

Table 5.19 Explainable
properties of
prototypes/criticisms

Properties	Values
Local or global	Global
Linear or non-linear	Non-linear
Monotonic or non-monotonic	Non-monotonic
Feature interactions captured	Yes
Model complexity	Medium

5.4.6 Influential Instances

When training a model, there are data instances that lie close the decision boundary. These instances have a substantial influence on the prediction or parameter determination of the model. There may also be outlier instances that may have an oversized effect on model training. Influential instances are the set of training data instances that influence the model such that their presence or absence can significantly change the model.

Influential instances can be identified by ablation. If removing an instance changes the model parameters or prediction, it is influential. Instead of focusing on how the features of an instance affect the model, we are focusing on how certain instances affect the model. Identifying influential instances can give us information about the robustness of a model or understanding of the data distribution.

A simple measure of influence I^{-i} on model prediction by the removal of data instance i is given by the modified Cook's distance (see Eq. (3.25)):

$$D_i = \sum_{j=1}^{n} (\hat{y}_j - \hat{y}_{j(i)})^2 \tag{5.61}$$

which measures the influence of an instance on the model prediction. Influential instances can also be identified using influence functions. Let θ be a model parameter vector and $\hat{\theta}_{\epsilon,z}$ be the vector by increasing the weight of an instance of interest z by a small amount ϵ:

$$\hat{\theta}_{\epsilon,z} = \arg\min_{\theta \in \Theta} (1 - \epsilon) \frac{1}{n} \sum_{i=1}^{n} L(z_i, \theta) + \epsilon L(z, \theta) \tag{5.62}$$

Here, $L(z_i, \theta)$ is the loss function for the model trained over a dataset D of n samples. The influence function of the model parameters is given by

$$I_{\text{up,params}}(z) = \left. \frac{d\hat{\theta}_{\epsilon,z}}{d\epsilon} \right|_{\epsilon=0} \tag{5.63}$$

and the influence function on the model prediction is given by

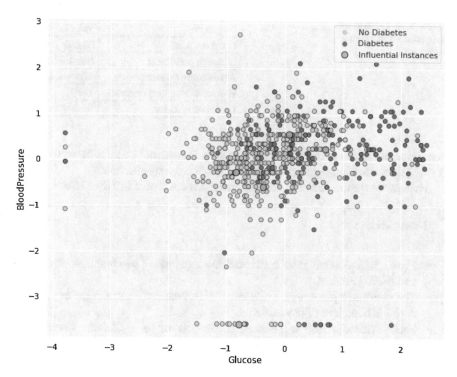

Fig. 5.25 Influential instances plotted along two features (*Glucose* and *BloodPressure*) in the Pima Diabetes dataset

$$I_{up,loss}(z, z_{test}) = \left. \frac{dL(z_{test}, \hat{\theta}_{\epsilon,z})}{d\epsilon} \right|_{\epsilon=0} \tag{5.64}$$

where z_{test} is the test data instances on which we evaluate the model.

Interpretation of Influential Instances: similar to LOCO for influential features, this method identifies the instances in the dataset that have the most impact on model prediction by deleting them from the training set and measuring changes in model error. These influential instances typically lie close to the decision boundary and help explain the global behavior of a model. Because the model will be retrained as many times as instances in the dataset, this method is computationally very expensive.

The influential instance method is simple, based on sample deletion, and is model-agnostic. It can provide useful local and global explanations. However, it is only applicable to models with differentiable parameters. Since it requires model

Table 5.20 Explainable
properties of influential
instances

Properties	Values
Local or global	Global
Linear or non-linear	Non-linear
Monotonic or non-monotonic	Non-monotonic
Feature interactions captured	Yes
Model complexity	High

retraining, it can be computationally expensive. Furthermore, as it is based on single sample deletion, sample group interaction is not properly accounted.

Explainable properties of Influential Instances are shown in Table 5.20.

Observations:

- Figure 5.25 plots the most influential instances along two features: *Glucose* and *BloodPressure*.
- The non-linear placement of the influential instances illustrates the complex decision boundary of the model.
- Each influential instance is seen to lie near or on a decision boundary between two groups of opposite predictions.

References

[FRD19] A. Fisher, C. Rudin, F. Dominici, All models are wrong, but many are useful: learning a variable's importance by studying an entire class of prediction models simultaneously (2019). arXiv:1801.01489 [stat.ME]

[FP08] J.H. Friedman, B.E. Popescu, Predictive learning via rule ensembles. Ann. Appl. Stat. 2(3) (2008). ISSN: 1932-6157. http://dx.doi.org/10.1214/07-AOAS148

[LL17b] S. Lundberg, S.-I. Lee, A unified approach to interpreting model predictions (2017). arXiv:1705.07874 [cs.AI].

[RSG16a] M.T. Ribeiro, S. Singh, C. Guestrin, "Why should I trust you?": explaining the predictions of any classifier (2016). arXiv:1602.04938 [cs.LG]

[Sha53] L.S. Shapley, A value for n-person games, in *Contributions to the Theory of Games (AM-28), Volume II*, ed. by H.W. Kuhn, A.W. Tucker (Princeton University Press, Princeton, 1953). ISBN: 978-1-40088-197-0

[Sla+20] D. Slack et al., Fooling LIME and SHAP: adversarial attacks on post hoc explanation methods (2020). arXiv:1911.02508 [cs.LG]

[SN19] M. Sundararajan, A. Najmi, The many Shapley values for model explanation (2019). arxiv:1908.08474, Comment: 9 pages

Chapter 6
Explainable Deep Learning

Recent advances in deep learning have made tremendous progress in the adoption of neural network models for tasks from resource utilization to autonomous driving. Most deep learning models are opaque black-box models that are not easily explainable. Unlike linear models, the weights of a neural network are not inherently interpretable to humans. The need for explainable deep learning has led to the development of a variety of methods that can help us better understand the decisions and decision-making process of neural network models. We note that many of the general post-hoc model-agnostic methods presented in Chap. 5 are applicable to deep learning models. This chapter presents a collection of explanation approaches that are specifically developed for neural networks by leveraging architecture or learning method.

6.1 Applications

The need for explainable deep learning is being driven by many real-world needs and applications. We discuss three broad categories: model validation, debugging, and exploration.

1. **Model Validation**: Model validation is the task of assessing how well a model behaves as intended in the real world. By doing so, we can assess the effectiveness and accuracy of a model. Explainable AI techniques can help us understand when errors in prediction occur and why they occur.
2. **Model Debugging**: Sometimes, a model may behave as intended but possess hidden biases. In recent years, we have seen real-life consequences of data and model biases. Our understanding of the weaknesses and limitations of deep learning models is vital for their adoption. Model debugging through explainable AI can help us uncover these deficiencies and provide insights into solutions to overcome them.

© The Author(s), under exclusive license to Springer Nature Switzerland AG 2021
U. Kamath, J. Liu, *Explainable Artificial Intelligence: An Introduction to Interpretable Machine Learning*, https://doi.org/10.1007/978-3-030-83356-5_6

3. **Model Exploration**: With the exponential growth in complexity of recent deep learning models (for instance, GPT-3 consists of 96 layers and 175 billion parameters), we have arrived at a point where we are unsure of what hidden abilities these models possess. Model exploration is the task of assessing the performance of a model on tasks beyond what it was originally intended. Explainable AI has become essential in order for us to understand the capabilities of these deep models through model exploration.

6.2   Tools and Libraries

Table 6.1 provides details of all the libraries used for various models in the chapter.

Table 6.1 Models and implementations

Model/Algorithm	Library
Attention (NMT)	TensorFlow
Attention (image captioning)	TensorFlow
LIME	Captum (PyTorch)
Occlusion	Captum (PyTorch)
RISE	Keras
Activation Maximization	tf-keras-vis
Saliency map	Captum (PyTorch)
DeepLIFT	Captum (PyTorch)
DeepSHAP	Captum (PyTorch)
Deconvolution	Captum (PyTorch)
Guided Backprop	Captum (PyTorch)
Integrated Gradients	Captum (PyTorch)
Layer-wise relevance propagation	Captum-0.4.0 (PyTorch)
Excitation backpropagation	excitationbp (PyTorch)
GradCAM	Captum (PyTorch)
TCAV	Captum (PyTorch)

6.3   Intrinsic

Intrinsic explainable deep learning methods leverage inherent model architecture to provide explanations of model predictions. Intrinsic methods encompass two types: attention-based and jointly trained multi-task models. One distinction between intrinsic and post-hoc explainable deep learning methods is that they can provide explanations even during training.

6.3.1   *Attention*

Attention-based neural networks mimic cognitive attention processes and can implicitly provide explanations of their output directly from the weights of their attention mechanism. The weights of the attention layer are learned during training. These weights inform which parts of the input feature space are "attended to" and influence the prediction. They provide a measure of feature importance and can be visualized via heatmaps, such as Fig. 6.1 for a word model or Fig. 6.2 for a visual classifier.

Attention mechanisms can provide useful feature-based explanations, but they are limited by the scope of the input space. Furthermore, they are interpretable only if the inputs are themselves interpretable. For intermediate representations in higher layers of a deep neural network, interpretability may be difficult. Serrato and Smith [SS19] have shown that attention weights may not necessarily correspond to importance or represent optimal explanations.

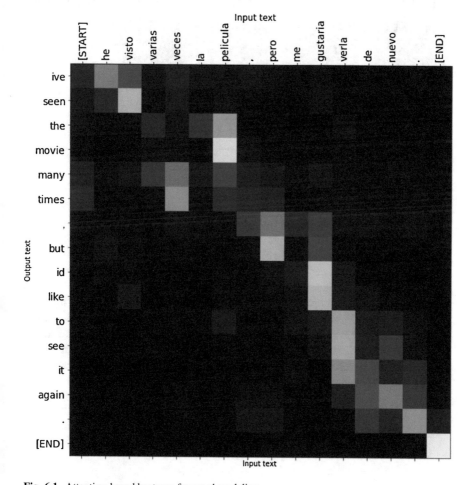

Fig. 6.1 Attention-based heatmap for word modeling

Interpretation of Attention Mechanisms: The attention weights of an attention-based model provide an intrinsic explanation to feature relevance on model predictions, but these weights may not provide optimal or interpretable explanations.

Attention-based neural networks are useful for tasks in NLP (RNNs and LSTMs), computer vision (CNNs), classification, and others.

Explainable properties of Attention-based methods are shown in Table 6.2.

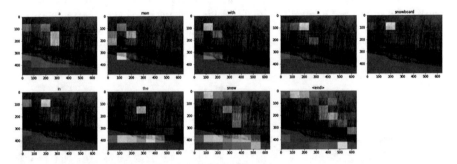

Fig. 6.2 Attention-based heatmap for visual classifier

Observations:

- Fig. 6.1 is a visualization of the attention weights for a TensorFlow Neural Machine Translation (NMT) model based on Bahdanau attention. The bi-LSTM model was trained on the Anki Spanish-to-English dataset for 5 epochs. Lighter colors indicate stronger weights. Word alignment (e.g., *movie ⇔ pelicula, seen ⇔ visto*) explanations are directly observable from the attention weights.
- Fig. 6.2 shows a sequence of blended attention masks for a neural image captioning system with a visual attention mechanism that shows what parts of the image the model focuses on as it generates a caption. The TensorFlow CNN with soft attention model was trained on the MS-COCO dataset for just 10 epochs. Note that the visual attention mechanism correctly attended to "man," but incorrectly attended to "snowboard."

Table 6.2 Explainable properties of attention-based methods

Properties	Values
Local or global	Local
Linear or non-Linear	Non-linear
Monotonic or non-monotonic	Non-monotonic
Feature interactions captured	Yes
Model complexity	Low

6.3.2 Joint Training

Another intrinsic method is to adopt a multi-task approach where an additional task is jointly trained to provide model explanations. This task can be trained to provide text-based justifications [LYW19, Zel+19], heatmaps over the feature [LBJ16, Iye+18] or concept space [AJ18, Don+17], or model prototypes [Li+17, Che+19]. Figure 6.3 illustrates an example of how a DNN architecture can be augmented with an explanation task that is jointly trained.

Joint training is a flexible method that enables high quality explanations at the cost of computational complexity due to changes in model architecture. The explanation task often comes a greater need for more data, and it may require explanation annotations in the training data in order to be trained with supervision.

Explainable properties of Joint-Training methods are shown in Table 6.3.

Interpretation of Joint Training Tasks: Augmenting network architecture to jointly train an explanation task is very flexible and can provide high quality explanations but can impose a heavy computational burden and requires training data with explanation annotations.

Observations:

- High quality explanations can be generated by jointly training a classifier against a gold set of explanations, but it can add significant complexity and computational cost.
- Explanation quality can be evaluated by calculating an explanation factor based on the explanation classifier output and target model output.

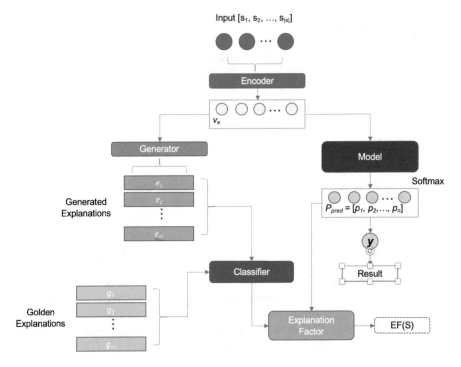

Fig. 6.3 Joint training for explanations

Table 6.3 Explainable
properties of joint-training
methods

Properties	Values
Local or global	Both
Linear or non-Linear	Both
Monotonic or non-monotonic	Both
Feature interactions captured	Yes
Model complexity	High

6.4   Perturbation

Perturbation methods attempt to explain feature relevance by measuring changes
in prediction score as features are altered. Perturbations at a feature level include
replacing, omitting individual features or groups, and learning attribution masks
that can explain the contributions of features.

6.4.1 LIME

As noted in Chap. 5, surrogate explanation methods replace complex models with simpler models that approximate the predictions of the original model. For neural networks, this process is known as "model distillation" where the knowledge encoded in a neural net is distilled into an interpretable machine learning model that can mimic its behavior [HVD15]. Explanation of the original neural network is provided through this interpretable model.

Local Interpretable Model-agnostic Explanations (LIME) is a useful method for generating local explanations of a model for specific instances. As LIME is model-agnostic, it is applicable to a variety of neural networks. LIME maps input data to an interpretable representation $x \rightarrow z = g(x)$, which is typically a binary vector used to represent the presence or absence of specific features in the input. For images or text, this could be the presence or absence of a patch of pixels or a set of words, respectively. It seeks to learn an interpretable model $h(z)$ by optimizing with the objective

$$\arg\min_{g \in \mathcal{G}} \mathcal{L}(f, g, \pi_x) + \Omega(g) \tag{6.1}$$

where π_x is a distance penalty between samples z and x, $\mathcal{L}$ is a measure of the unfaithfulness of g in imitating f in the local region defined by π_x, and Ω is a complexity penalty that ensures the learned model is not too complex. As an example, we can generate local explanations for a neural network image classifier by applying the following:

$$h(z) = a_g^T x$$

$$\pi_x(z) = \exp\left(-||x - z||^2/\sigma^2\right)$$

$$\mathcal{L}(f, g, \pi_x) = \sum_{z,z'} \pi_x(z) \left(f(z) - h(z')\right)^2$$

$$\Omega(h(z)) = ||a_g||$$

As previously noted, LIME explanations require a large number of randomly perturbed samples to compute accurate local explanations of a complex model. The class of each of these samples must be first predicted by a forward pass of the complex model, which could add computational burden when explaining large neural networks.

Explainable properties of LIME are shown in Table 6.4.

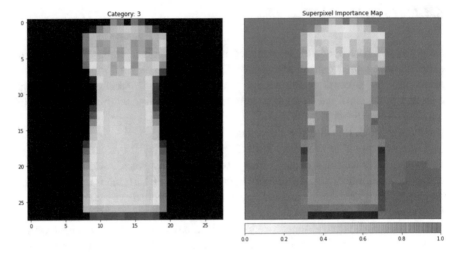

Fig. 6.4 LIME explanations on Fashion MNIST

Observations:

- The left side of Fig. 6.4 shows the model prediction from a PyTorch CNN model with three convolutional layers and two fully connected layers trained on Fashion MNIST for 2 epochs.
- LIME provides local explanations for image classification by computing the effect of the presence or absence of superpixels on the classification.
- The right side shows the importance of the superpixels (of a single color) on the prediction category 3 (Dress).

Table 6.4 Explainable properties of LIME

Properties	Values
Local or global	Local
Linear or non-linear	Both
Monotonic or non-monotonic	Both
Feature interactions captured	Yes
Model complexity	Medium

6.4.2 *Occlusion*

Perhaps one of the easiest ways to perturb an instance, occlusion (also named feature ablation) is a local method by which the input features of an instance are sequentially replaced with a constant (commonly zero). In 2013, [ZF13] proposed occlusion sensitivity as an explanation method for image classification by systematically occluding different portions of the input image with a gray patch sliding window. Feature relevance is measured by the change in prediction accuracy of the correct class or feature activation magnitude of the last neural network layer. This approach is applicable to other machine learning tasks as well. In 2017, [LMJ17] proposed an occlusion method for natural language-related tasks termed representation erasure, where input words are systematically erased to determine their contribution to prediction accuracy.

> Interpretation of Occlusion: Occlusion methods are local explanations similar to ablation methods in Chap. 5, where input features are systematically replaced with a constant. The more prediction accuracy drops, the more significant the occluded features. Occlusion methods are computationally efficient but do not capture feature interaction well.

Like the feature ablation methods of Chap. 5, occlusion has limited ability to capture feature interaction effects. If interaction effects are significant, occlusion will likely return incorrect results. Unlike LOCO, occlusion is a local method that is easy to compute and does not require model retraining.

Explainable properties of Occlusion are shown in Table 6.5.

Observations:

- The left side of Fig. 6.5 shows two category 9 (shoe) classifier predictions by a PyTorch 3-convolutional layer CNN model trained on Fashion MNIST.
- The right side of Fig. 6.5 shows the occlusion importance maps generated by sliding a black 3x3 pixel mask across the image and measuring the resulting change in prediction probability.
- For both images, the importance maps indicate the diagonal area which most influence the classifier prediction for category 9.

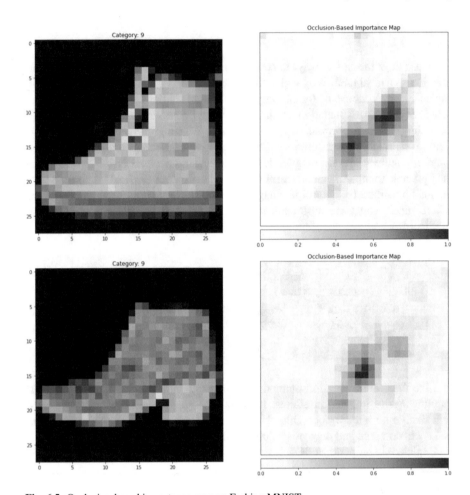

Fig. 6.5 Occlusion-based importance map on Fashion MNIST

Table 6.5 Explainable
properties of occlusion

Properties	Values
Local or global	Local
Linear or non-linear	Non-linear
Monotonic or non-monotonic	Non-monotonic
Feature interactions captured	No
Model complexity	Low

6.4.3 RISE

In 2018, [PDS18a] proposed the Random Input Sampling for Explanations (RISE)
method as generalized version of occlusion by probing a model with randomly
masked portions of the input instance. Given a random mask M, an input instance

x, and a model $f(x)$, the feature importance of x_j (the j-th feature of x) is given by

$$S_f(x_j) = \mathbb{E}_M\left[f(x \star M)|M(x_j) = 1\right] \qquad (6.2)$$

where $\star$ denotes element-wise multiplication. Thus, RISE computes the importance map as the weighted average of random masks.

In practice, Monte Carlo sampling is used to generate random masks to compute RISE. As binary masks suffer when feature interactions exist, [PDS18a] proposed a soft version that up-samples a small binary mask using bilinear interpolation. The resulting mask values are continuous across [0, 1].

Explainable properties of RISE are shown in Table 6.6.

Interpretation of RISE: As a Monte Carlo sampled occlusion method, RISE is useful for generating importance map explanations of specific instances. By incorporating bilinear interpolation of smaller binary patches, RISE can take feature interactions into account.

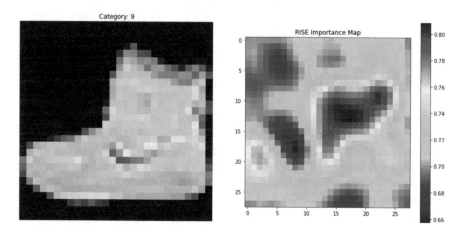

Fig. 6.6 RISE importance map on Fashion MNIST

Observations:

- The left image in Fig. 6.6 was classified as category 9 (shoe) by a 2-layer Keras CNN model trained on Fashion MNIST.
- The RISE importance map on the right side was computed using 2000 Monte Carlo generated random pixel maps.
- The right image shows higher importance around the ankle and toe area of the shoe image, while little to no importance with the rest of the image.
- RISE provides better importance maps than occlusion but at higher computation cost.

Table 6.6 Explainable properties of RISE

Properties	Values
Local or global	Local
Linear or non-linear	Non-linear
Monotonic or non-monotonic	Non-monotonic
Feature interactions captured	Yes
Model complexity	Medium

6.4.4 Prediction Difference Analysis

Prediction Difference Analysis (PDA) was proposed by [Zin+17] as a classifier explanation method to assign a relevance value to each input feature with respect to each class. It estimates this relevance by measuring how prediction changes if a feature value is unknown. Since for neural networks it is impractical to either label a feature as unknown or retrain the model with the feature left out (e.g., LOCO), PDA simulates the absence of the feature by marginalizing over it. Given a class c and an input instance x with j-th feature x_j, the class probability with unknown feature x_j is given by

$$p(c|x_{-j}) = \sum_{x_j} p(x_j|x_{-j})p(c|x_{-j}, x_j) \tag{6.3}$$

where x_{-j} is the set of all features in x except the j-th feature and the summation is taken over all possible values of x_j. For large feature spaces, computational efficiency can be gained by assuming feature x_j is uncorrelated with the other features x_{-j}, and the class probability becomes

$$p()c|x_{-j}) \approx \sum_{x_j} p(x_j)p(c|x_{-j}, x_j)$$ (6.4)

where $p(x_j)$ can be approximated by its empirical distribution. PDA compares the class probability with all features present $p(c|x)$ with $p(c|x_{-j})$ to determine feature relevance by defining a weight-of-evidence function:

$$WE_j(c|x) = \log_2\left(odds(c|x)\right) - \log_2\left(odds(c|x_{-j})\right)$$ (6.5)

where

$$odds(c|x) = \frac{p(c|x)}{1 - p(c|x)} \quad odds(c|x_{-j}) = \frac{p(c|x_{-j})}{1 - p(c|x_{-j})}$$ (6.6)

To account for zero probabilities, [Zin+17] proposed using a Laplace correction to the class probability:

$$p(c|x) \leftarrow \frac{p(c|x)n + 1}{n + k}$$ (6.7)

where N is the number of training instances and k is the number of classes. The magnitude of the evidence function WE_j indicates the significance of the j-th feature on class c prediction. A positive value of WE_j implies that feature x_j contributed positively to the evidence for class c, and removing it would reduce confidence in prediction for the class. A negative value implies evidence against the class.

When feature interactions exist, PDA can be adjusted to account for neighbor interactions. Instead of assuming feature x_j is uncorrelated with every other feature and replacing the conditional probability $p(x_j|x_{-j})$ with $p(x_j)$, [Zin+17] proposed using conditional sampling where a neighborhood patch of features around and including x_j is marginalized:

$$p(x_j|x_{-j}) \approx p(x_j|\hat{x}_{-j})$$ (6.8)

Here, $\hat{x}_{-j}$ is the set of all features except for the patch of features around and including x_j.

PDA is also applicable for visualizing neuron contributions to hidden layer activations. Given a hidden layer H with neuron values h and the i-th neuron in the subsequent layer that depends on H with value $z_i(h)$, the activation function g when the j-th neuron value in H is unknown is given by

$$g(z_j|h_{-j}) = \sum_{h_j} p(h_j|h_{-j})z_i(h_{-j}, h_j)$$ (6.9)

and the activation difference AD_j is a measure of the contribution of the j-th neuron in hidden layer H to the i-th neuron in the subsequent layer:

$$AD_j(z_i|h) = g(z_i|h) - g(z_i|h_{-j}) \qquad (6.10)$$

In practice, PDA is fairly computationally intensive, especially if conditional sampling is used to capture neighbor feature interactions.

Explainable properties of Prediction Difference Analysis are shown in Table 6.7.

> Interpretation of Prediction Difference Analysis: PDA is a local method that measures feature relevance by taking the class prediction probability differences while marginalizing over a feature or a patch of features. It can account for feature interactions but comes with a larger computation cost.

Table 6.7 Explainable properties of prediction difference analysis

Properties	Values
Local or global	Local
Linear or non-linear	Non-linear
Monotonic or non-monotonic	Non-monotonic
Feature interactions captured	Yes
Model complexity	Medium

6.4.5 Meaningful Perturbation

Meaningful perturbation (MP) [FV17] is a local explanation method based on a framework of meta-predictors to explain predictions for neural classifiers. These meta-predictors are trained to predict the presence or absence of input features. Their prediction error is a measure of the faithfulness of the explanation.

For a given instance x_0, the method applies a set of meaningful, local perturbations given by

$$[\Phi(x_0, m)] = \begin{cases} m(u)x_0(u) + (1 - m(u))\mu_0, & constant \\ m(u)x_0(u) + (1 - m(u))\eta(u) & noise \\ \int g_{\sigma_0 m(u)}(v - u)x_0(v)dv & blur \end{cases} \qquad (6.11)$$

where μ_0 is the average color, $\eta(u)$ are i.i.d. Gaussian noise samples, and σ_0 is the standard deviation of the Gaussian blur kernel g_σ. The method plays a "deletion game," which seeks to find the smallest deletion mask m^* that causes the classifier score f_c for class c to drop $f_c(\Phi(x_0, m)) < fc(x_0)$ by optimizing:

$$m^* = \underset{m \in [0,1]^d}{\arg\min} \lambda||1 - m||_1 + f_c(\Phi(x_0, m)) \qquad (6.12)$$

where d is the total number of features and ϵ is a hyperparameter. A symmetric "preservation game" can also be played, which seeks to find the smallest subset of the image that must be retained to preserve the classifier score $f_c(\Phi(x_0, m)) \geq fc(x_0)$ by optimizing:

$$m^* = \underset{m \in [0,1]^d}{\arg\min} \lambda ||m||_1 + f_c(\Phi(x_0, m)) \tag{6.13}$$

The deletion game tries to remove just enough evidence to prevent the model from recognizing the class, while the preservation game ties to keep just enough evidence. Both of these optimizations can be solved by gradient descent.

To mitigate the effects of artifacts that might exist in the trained neural network, meaningful perturbations propose a modified deletion game where the learned mask is regularized:

$$m^* = \min_{m \in [0,1]^d} \lambda_1 ||1 - m||_1 + \lambda_2 \sum_u ||\nabla m(u)||_\beta^\beta + \mathbb{E}_\tau \left[f_c(\Phi(x_0(\cdot - \tau), m)) \right] \tag{6.14}$$

This optimization can be solved with stochastic gradient descent.

> Interpretation of Meaningful Perturbations: MP is a global method that learns where a neural classifier looks by discovering features that most affect its class prediction output score when locally perturbed. It learns a feature mask that explains the classification result as an optimization problem.

In practice, the algorithm learns the smallest, low-resolution, sparse set of masks, which, when up-sampled and added to the input instance, causes the target class prediction to drop.

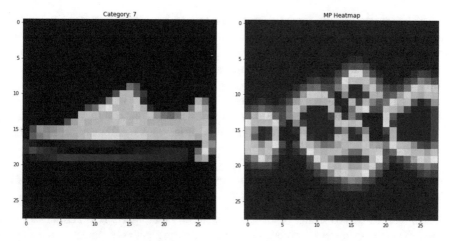

Fig. 6.7 Meaningful perturbations mask on Fashion MNIST

Explainable properties of Meaningful Perturbation are shown in Table 6.8.

Observations:

- The left image in Fig. 6.7 shows the category predictions of a PyTorch 3-layer CNN model trained on Fashion MNIST.
- The right image shows the meaningful perturbations heatmap for the prediction of category 7 (Sneaker).

Table 6.8 Explainable properties of meaningful perturbation

Properties	Values
Local or global	Local
Linear or non-linear	Non-linear
Monotonic or non-monotonic	Non-monotonic
Feature interactions captured	Yes
Model complexity	Medium

6.5 Gradient/Backpropagation

Whereas perturbation-based explanation methods leverage variations in input features to explain feature relevance, gradient methods leverage the flow of information during backpropagation to explain the relationship between input features and network output. Gradient-based methods typically provide visual explanations through heatmaps of neuron or feature attributions.

6.5.1 Activation Maximization

Visual explanations provide an efficient and human-interpretable method to understand deep neural network predictions. One of the earliest global explanation methods is the Activation Maximization method [ECB10], which visually identifies the input features that can create the greatest response in the output of specific neurons.

Given a neural network with parameters θ, an input sample x, and the i-th neuron in the j-th layer with activation $h_{ij}(\theta, x)$, the goal of is to find a hypothetical x^* that can maximize the activation of this neuron. This can be expressed as the optimization

$$x^* = \arg\max_{x} h_{ij}(\theta, x) \tag{6.15}$$

which can be solved using gradient ascent in the input space. It is similar to the backpropagation method, except instead of adjusting network parameters θ, the optimization is over the input space while the network parameters are held constant. The synthetic instance x^* can be visualized and will represent the input feature pattern that will maximize the activation of a specific neuron in the network.

> Interpretation of Activation Maximization: AM is a global method that finds the input pattern that can generate the highest activation in the response of a specific neuron in a deep neural network.

Given the non-linear activations, there are no guarantees that gradient ascent will identify a unique global optimum x^*, but in practice using multiple random starting points and either averaging or selecting the maximum activation has been shown to be effective.

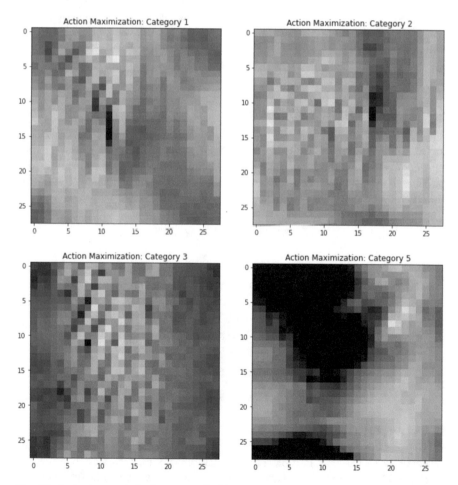

Fig. 6.8 Activation maximization map on Fashion MNIST

Explainable properties of Activation Maximization are shown in Table 6.9.

Observations:

- Fig. 6.8 shows the activation maximization maps for 4 neurons at the output layer of a Keras 2-layer CNN model trained on Fashion MNIST. These neurons correspond to 4 classification categories.
- Note that it is difficult to determine the original classification category from the activation maps (1=Trouser, 2=Pullover, 3=Dress, 5=Sandal).

Table 6.9 Explainable properties of activation maximization

Properties	Values
Local or global	Global
Linear or non-Linear	Non-linear
Monotonic or non-monotonic	Non-monotonic
Feature interactions captured	Yes
Model complexity	Low

6.5.2 Class Model Visualization

Activation maximization is the basis of class model visualizations. Given a neural network classifier with scoring function $S_c(x)$ for an output class c and input x, it is possible to learn an instance x' that is most representative of the class by optimizing the equation

$$x' \leftarrow \arg\max_x S_c(x) - \lambda||x||_2 \qquad (6.16)$$

where λ is a regularization parameter. The generated instances for each class are learned representations by the neural network and can be very visually entertaining.

> Interpretation of Class Model Visualization: this global explanation method learns the input patterns that generate the greatest activation for a specific model class. When visualized, these patterns can provide colorful explanations of what the model has learned.

Class model visualizations have recently gained widespread attention for pioneering a new branch of deep learning-generated art called "deep dream" and "Inceptionism" based on the colorful visualizations of model classes.

Explainable properties of Class Model Visualization are shown in Table 6.10.

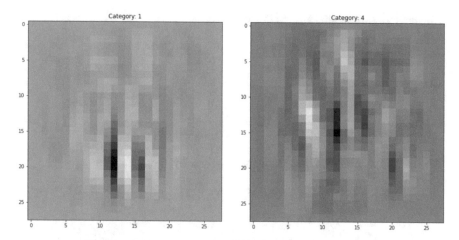

Fig. 6.9 Class model visualization on Fashion MNIST

Observations:

- Fig. 6.9 shows the model visualizations for category 1 (Trouser) and category 4 (Coat) from a PyTorch 3-layer CNN model trained on Fashion MNIST.
- When viewed at a distance, the left image hints at a pair of trousers, while the right image gives the semblance of a coat.

Table 6.10 Explainable properties of class model visualization

Properties	Values
Local or global	Global
Linear or non-linear	Non-linear
Monotonic or non-monotonic	Non-monotonic
Feature interactions captured	Yes
Model complexity	Low

6.5.3 Saliency Maps

Saliency maps [SVZ14a] provide local explanations for specific instances. Given an instance of interest x_0, we can approximate the non-linear scoring function $S_c(x)$ by a Taylor series expansion around this instance:

$$S_c(x) \approx w^T x + b \qquad (6.17)$$

where w are the saliency weights

$$w = \left. \frac{\partial S_c}{\partial x} \right|_{x=x_0} \qquad (6.18)$$

The saliency map M_j for the j-th feature is given by

$$M_j = |w_j| \qquad (6.19)$$

> Interpretation of Instance-Specific Saliency Maps: Instance-specific saliency maps are a local explanation method that takes the partial derivative of the scoring function with respect to each feature as a measure of feature importance. They are very quick to calculate but require the scoring function to be differentiable.

Instance-specific class saliency maps are extremely quick to compute and do not require any additional annotation to provide explanations. They do, however, require the scoring function to be differentiable.

Explainable properties of Saliency Maps are shown in Table 6.11.

Observations:

- The left side of Fig. 6.10 shows the category predictions of a PyTorch 3-layer CNN model trained on Fashion MNIST.
- The right side shows the saliency maps for an image of category 0 (t-shirt) and category 7 (sandal).

Table 6.11 Explainable properties of saliency maps

Properties	Values
Local or global	Local
Linear or non-linear	Linear
Monotonic or non-monotonic	Monotonic
Feature interactions captured	No
Model complexity	Low

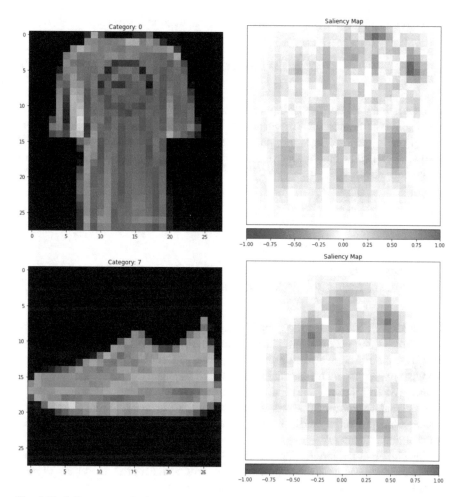

Fig. 6.10 Saliency maps for instances of Fashion MNIST

6.5.4   DeepLIFT

Deep Learning Important Features (DeepLIFT) [SGK19] is a recursive, local explanation method that decomposes a neural network model prediction for a specific instance by backpropagating the contributions of the neurons through the network. DeepLIFT is based on the difference between the activation of each neuron and its "reference activation" in order to compute contribution scores. This reference activation represents a default or a neutral input.

Consider a neuron of interest with activation $f(x)$ and with a set of input neurons $x_1, x_2, \ldots, x_n$. If $f(x')$ represents the reference activation of the neuron of interest for a reference input x', the difference in neural activation is given by

$$\Delta t = f(x) - f(x') = \sum_{i=1}^{n} C_{\Delta x_i \Delta t} \qquad (6.20)$$

where n is the total number of input neurons necessary to compute $f(x)$ and $\Delta x = x - x'$. This is termed the "summation-to-delta" property. The contribution score $C_{\Delta x_i \Delta t}$ relates how changes in input Δx affect changes in neuron activation Δt. If we divide this contribution score by Δx, we can define a multiplier analogous to a partial derivative:

$$m_{\Delta x \Delta t} = \frac{C_{\Delta x \Delta t}}{\Delta x} \qquad (6.21)$$

This multiplier follows a useful chain rule:

$$m_{\Delta x_i \Delta t} = \sum_{j} m_{\Delta x_i \Delta y_j} m_{\Delta y_j \Delta t} \qquad (6.22)$$

This chain rule allows for the contribution scores to be backpropagated layer-by-layer through the network and is analogous to how gradients are backpropagated. By using the difference from reference approach, DeepLIFT allows contribution scores to propagate even when the gradient is zero.

DeepLIFT proposes three contribution scoring functions: a linear rule applicable to dense and convolutional layers, a rescale rule that can account for saturation and thresholding problems, and a reveal-cancel rule that treats positive and negative contributions separately.

The choice of a reference input x' is an important consideration, as it determines what relevance scores are computed against. For instance, the use of an all zero reference may not be as useful if noise is present in the background.

Interpretation of DeepLIFT: as a local explanation method, DeepLIFT calculates input importance relative to a reference by backpropagating contribution scores through the network. It is very computationally efficient and provides an approximation to Shapley values. The choice of scoring function and reference input should be carefully considered.

DeepLIFT scores can be efficiently computed with a single backward pass. They are connected to Shapely values, which measure the marginal contribution of each feature averaged across the set of all possible coalitions of features. If excluding a feature is equivalent to setting it to its reference value, DeepLIFT can be thought of as a fast approximation of the Shapely values.

Explainable properties of DeepLIFT are shown in Table 6.12.

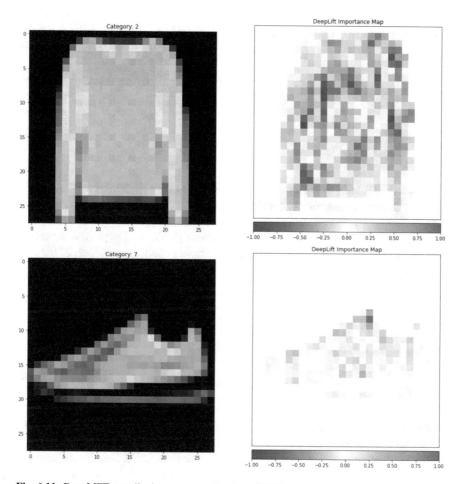

Fig. 6.11 DeepLIFT contribution scores on Fashion MNIST

Observations:

- The left side of Fig. 6.11 shows the category predictions of a PyTorch 3-layer CNN model trained on Fashion MNIST.
- The right side shows the DeepLIFT importance map based on the linear-rule and black baseline reference.
- Note that DeepLIFT provides better visual explanations in comparison to saliency maps.

Table 6.12 Explainable
properties of DeepLIFT

Properties	Values
Local or global	Local
Linear or non-linear	Non-linear
Monotonic or non-monotonic	Non-monotonic
Feature interactions captured	Yes
Model complexity	Low

6.5.5 DeepSHAP

DeepSHAP [CLL19] is an extension of the KernelSHAP method of Chap. 5 by leveraging the compositional architecture of deep neural networks to improve computational efficiency. As previously mentioned, the per node attribution rules in DeepLIFT can approximate the Shapley values [SGK19]. DeepSHAP leverages this approximation as well as DeepLIFT's multiplier chain rule. In DeepSHAP, the multipliers are expressed in terms of SHAP values ϕ_i:

$$m_{x_j, f_j} = \frac{\phi_i(f_j, x)}{x_j - \mathbb{E}[x_j]} \tag{6.23}$$

and follow the chain rule:

$$m_{x_j, f_j} = \sum_j m_{x_j, y_j} m_{y_j, f_j} \tag{6.24}$$

DeepSHAP calculates SHAP values for large networks by starting with Shapley values for simple network components and backpropagating them using this rule.

Rather than setting the reference input x' as in DeepLIFT, DeepSHAP approximates the reference value by averaging over background dataset instances. It can estimate approximate SHAP values such that they sum up to the difference between the expected model output on background instances and the current model output $f(x) - \mathbb{E}[f(x)]$.

> Interpretation of DeepSHAP: as a local explanation method, DeepSHAP is an extension of DeepLIFT to backpropagate Shapley values through the network. DeepSHAP computes an input reference as the expectation over background data instances. It is very computationally efficient and provides a quick approximation to Shapley values, which may be biased when features are strongly correlated.

DeepSHAP is computationally very efficient. Instead of depending on DeepLIFT's contribution rules to linearize each node in the network, DeepSHAP effectively linearizes the network by computing SHAP values using the chain rule.

As such, they are approximations to the true Shapley values and will be biased when strong feature interactions exist.

Explainable properties of DeepSHAP are shown in Table 6.13.

Observations:

- The left side of Fig. 6.12 shows the predictions of a PyTorch 3-layer CNN model trained on Fashion MNIST.
- The right side shows the DeepSHAP importance map based on a baseline reference consisting of a random sample of 10 images from the training set.
- Note that DeepSHAP provides slightly worse visual explanations in comparison to DeepLIFT but is more computationally efficient.

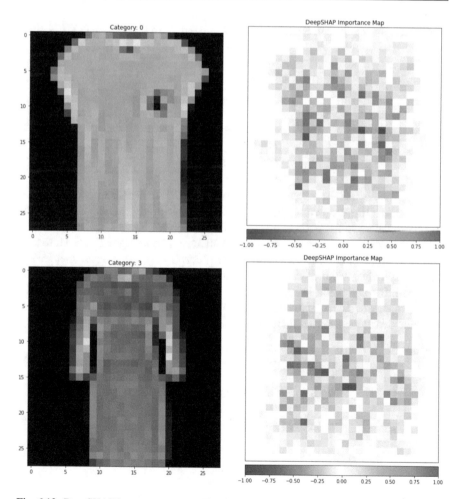

Fig. 6.12 DeepSHAP importance map on Fashion MNIST

Table 6.13 Explainable
properties of DeepSHAP

Properties	Values
Local or global	Local
Linear or non-linear	Linear
Monotonic or non-monotonic	Monotonic
Feature interactions captured	No
Model complexity	Low

6.5.6 Deconvolution

Deconvolution [ZF13] is an explanation method proposed for visualizing the feature
contributions of CNN architectures (convnets). It takes the output of the CNN and
runs the CNN in reverse. The guiding notion is to determine which portions of an
input instance are most discriminative for a single neuron.

Given a CNN neural network of J layers, the output of the j-th layer C^j is given
by

$$A^j = C^{j-1} * K^j + b^j \tag{6.25}$$

$$B^j = ReLU(A^j) = \max\left(A^j, 0\right) \tag{6.26}$$

$$C^j = \text{maxpool}\left(B^j\right) \tag{6.27}$$

$$s^j = \text{switch}\left(B^j\right) \tag{6.28}$$

where K^j and b^j are the learned filter and bias for the j-th layer, respectively. ReLU
is the rectified linear operator, and the switch variable s^j records the indices of the
maximum values in the pooling operation for the deconvolution step.

A deconvolution network (deconvnet) is attached to the original convnet to map
feature activations back to the input space. Each layer in this deconvnet inverts
the corresponding layer of the original convnet. To examine each convnet neuron
activation, the activation is set to zero for all other neurons in the layer, and the
feature maps are fed as input to the deconvnet layer, which sequentially applies
unpooling, rectification, and filtering operations.

$$\hat{C}^j = \text{unpool}\left(C^j, s^j\right) \tag{6.29}$$

$$\hat{B}^j = ReLU(\hat{C}^j) = \max\left(\hat{C}^j, 0\right) \tag{6.30}$$

$$\hat{A}^j = \left(\hat{B}^j - b^j\right) * K^{jT} \tag{6.31}$$

where K^{jT} is the transpose of K^j. Together, the set of deconvolution operations is called transpose convolution. While the max pooling operation is not invertible, unpooling can be performed if switch variables are recorded during the forward propagation of the convnet. The ReLU ensures feature maps are non-negative, and the filtering operation up-weights and up-scales the feature representation in each layer.

> Interpretation of Deconvolution: deconvolution explains learned feature maps in CNN-based models by propagating them through an inverted convolutional network called a deconvnet.

The deconvolution method is specific to CNN architectures, though the process is applicable to dense layers as well and other non-linearities beyond ReLU. In order to operate effectively, deconvolution requires a forward pass through the original convnet in order to calculate and store the switch variables to allow the deconvnet to reverse the max pooling operations. As a result, visualizations derived from deconvolution are conditioned on the instance used to calculate the switch variables and do not directly visualize the learned features [Spr+15b].

Explainable properties of Deconvolution are shown in Table 6.14.

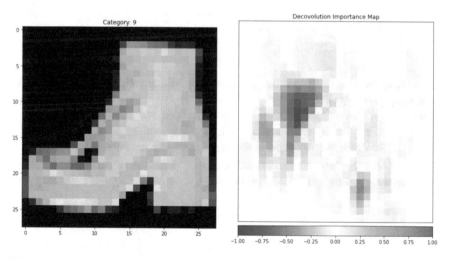

Fig. 6.13 Deconvolution visualizations on Fashion MNIST

Observations:

- The left side of Fig. 6.13 shows the input image and category prediction of a PyTorch 3-layer CNN model trained on Fashion MNIST.
- The right side shows the deconvolution feature importance map conditioned on the input image instance.

6.5.7 Guided Backpropagation

Guided backpropagation [Spr+15b] is another explanation method for feature attribution on CNN-based architectures. It is similar to deconvolution, except it removes the unpooling operation and adopts a modified operation for the ReLU non-linearity.

Table 6.14 Explainable properties of deconvolution

Properties	Values
Local or global	Global
Linear or non-linear	Non-linear
Monotonic or non-monotonic	Non-monotonic
Feature interactions captured	Yes
Model complexity	Medium

In deconvolution, only positive gradients are backpropagated. In vanilla backpropagation, the gradients for only positive inputs are kept for each layer. Guided backpropagation incorporates both of these methods such that only positive gradients associated with positive input values are backpropagated. This stops the backward flow of negative gradients through the inverse network.

Let f_i^j and R_i^j be the i-th neuron input to and feature map of the j-th layer, respectively. Then for guided backpropagation, we have during the backward pass

$$R_i^j = (f_i^j > 0)(R_i^{j+1} > 0) \ R_i^{j+1} \tag{6.32}$$

This is in contrast to deconvolution, where the backward pass applies the operation after unpooling:

$$R_i^j = (R_i^{j+1} > 0) \ R_i^{j+1} \tag{6.33}$$

Because of the additional guidance signal in guided backpropagation, unpooling is unnecessary and does not require an initial forward pass through the convnet to compute and store switch variables. As a result, it is more computationally efficient. Explanations via guided backpropagation are not conditioned on any single instance as in deconvolution and provide more accurate explanations of feature activations. Explainable properties of Guided Backpropagation are shown in Table 6.15.

Interpretation of Guided Backpropagation: guided backprop is an improvement upon deconvolution to explain learned feature maps in CNN-based models. It replaces unpooling and ReLU operations with an operation allow only positive gradients associated with positive inputs to be backpropagated.

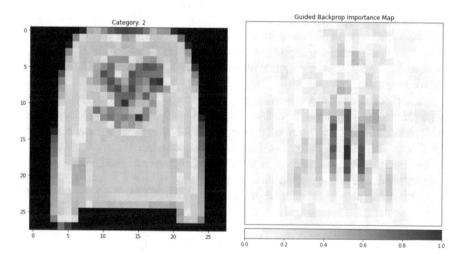

Fig. 6.14 Guided backpropagation visualizations on Fashion MNIST

Observations:

- The left side of Fig. 6.14 shows the category prediction of a PyTorch 3-layer CNN model trained on Fashion MNIST.
- The right side shows the guided backprop feature importance map without the need for a forward pass for conditioning as in deconvolution.
- Note the improvement over deconvolution in visual explanation.

Table 6.15 Explainable properties of guided backpropagation

Properties	Values
Local or global	Global
Linear or non-linear	Non-linear
Monotonic or non-monotonic	Non-monotonic
Feature interactions captured	Yes
Model complexity	Medium

6.5.8 Integrated Gradients

Integrated gradients [SN19] is an explanation method that does not require modification of the original network and takes an axiomatic approach to generate feature attributions for specific instances. Ideally, an attribution method should obey the axiom of sensitivity, which states that if two inputs x and x' differ by a single feature and have different prediction values $f(x) \neq f(x')$, then the feature should be given non-zero attribution. Unfortunately, many gradient-based attribution methods violate this sensitivity axiom, including DeepLIFT, deconvolution, and guided backpropagation.

An attribution method should also obey the axiom of implementation invariance, which states that two neural networks, even with vastly different implementations, are functionally equivalent if they map the same inputs to the same outputs. Attributions are identical across two functionally equivalent neural networks.

The integrated gradients method satisfies both of these axioms. Given a neural network model f, an input instance x, and a baseline reference x', one can traverse along the direct path from reference to input $x' \rightarrow x$ while accumulating gradients. The integrated gradient along the j-th feature is given by

$$IG_j(x) = (x_j - x_j') \int_{\alpha=0}^{1} \frac{\partial f \left(x' + \alpha \left[x - x' \right] \right)}{\partial x_j} \partial \alpha \qquad (6.34)$$

Integrated gradients satisfy the axiom of completeness, which states that the sum of the attributions is equal to $f(x) - f(x')$.

A key consideration with integrated gradients is selecting a baseline, similar to DeepLIFT. Ensuring that the baseline reference has near-zero score is important to ensure the attributions are derived from the input rather than the baseline.

In practice, the path integral is calculated using a Riemann summation approximation:

$$IG_j(x) \approx \frac{(x_j - x_j')}{M} \sum_{i=1}^{M} \frac{\partial f \left(x' + \frac{i}{M}(x - x') \right)}{\partial x_j} \qquad (6.35)$$

where the parameter M is the number of steps in the Riemann summation.

Interpretation of Integrated Gradients: IG takes an axiomatic approach to computing feature attributions by accumulating gradients along the direct path between a baseline reference and an instance of interest. They are simple and quick to compute.

The integrated gradients method is not limited to CNNs and can be applied to a wide variety of deep neural networks. They are computationally efficient and simple to compute.

Explainable properties of Integrated Gradients are shown in Table 6.16.

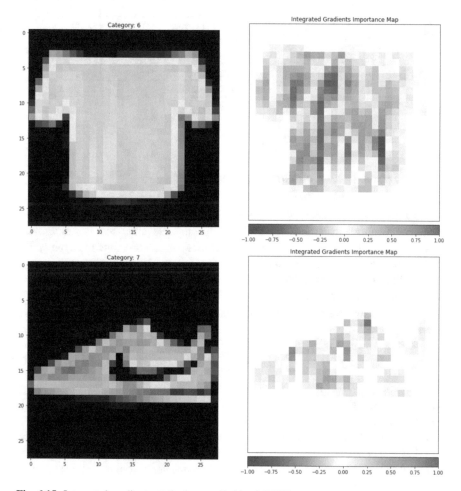

Fig. 6.15 Integrated gradients attributions on Fashion MNIST

Observations:

- The left side of Fig. 6.15 shows the category prediction of a PyTorch 3-layer CNN model trained on Fashion MNIST.
- The right side shows the integrated gradients importance map using a zero baseline and 50-step Riemann approximation.
- Note the quality of the visual explanations in comparison to DeepLIFT.

Table 6.16 Explainable properties of integrated gradients

Properties	Values
Local or global	Local
Linear or non-linear	Non-linear
Monotonic or non-monotonic	Non-monotonic
Feature interactions captured	Yes
Model complexity	Medium

6.5.9 Layer-Wise Relevance Propagation

Layer-wise relevance propagation (LRP) [Bac+15a] is an explanation method that computes feature attributions by backpropagating relevance scores through the network layers from output to input. Relevance scores measure the connection strength between any two neurons.

LRP follows the law of relevance conservation, which states that the relevance of any neuron is equal to the sum of its relevance maps in the previous layer. That is, if the relevance score for the i-th neuron in layer j is given by R_i^j, then conservation states that

$$f(x) = \sum_k R_k^{j+1} = \sum_i R_i^j = \ldots = \sum_m R_m^1 \qquad (6.36)$$

where $f(x)$ is the neural network prediction for input x. The total relevance is preserved across layers.

LRP starts with a forward pass on an instance of interest and the class prediction of a single neuron in the top layer with all other neurons at zero value. The relevance is set equal to this class prediction and is backpropagated through the network with the propagation rule:

$$R_j = \sum_k \frac{a_j w_{jk}}{\sum_0^j a_j w_{jk}} R_k \qquad (6.37)$$

where neurons j and k are in consecutive layers, a_j is the activation for the neuron in layer j, and w_{jk} is the connection weight between these two neurons. Relevance scores can thus be recursively calculated back to the input and then visualized as a heatmap to explain input feature attribution. Note that other propagation rules can be used for specific applications, and different rules can be used for different layers (Table 6.17) so long as the law of conservation is followed [Mon+19].

Explainable properties of Integrated Gradients are shown in Table 6.18.

> Interpretation of Layer-wise Relevance Propagation: LRP provides visual explanations of individual instances by backpropagating relevance scores from the neural network top layer down to the input. By construction, the total relevance is conserved across layers. The choice of propagation rule is a design consideration.

Table 6.17 List of commonly used LRP rules

Rule	Layer
$R_j = \sum_k \frac{a_j w_{jk}}{\sum_{0,j} a_j w_{jk}} R_k$	Upper
$R_j = \sum_k \frac{a_j w_{jk}}{\epsilon + \sum_{0,j} a_j w_{jk}} R_k$	Middle
$R_j = \sum_k \frac{a_j \left(w_{jk} + \gamma w_{jk}^+ \right)}{\sum_{0,j} a_j \left(w_{jk} + \gamma w_{jk}^+ \right)} R_k$	Lower
$R_j - \sum_k \left(\alpha \frac{(a_j w_{jk})^+}{\sum_{0,j} (a_j w_{jk})^+} - \beta \frac{(a_j w_{jk})^-}{\sum_{0,j} (a_j w_{jk})^-} \right) R_k$	Lower
$R_j = \sum_k \frac{1}{\sum_j 1} R_k$	Lower
$R_j = \sum_j \frac{w_{ij}^2}{\sum_i w_{ij}^2} R_j$	First
$R_i = \sum_j \frac{x_i w_{ij} - l_i w_{ij}^+ - h_i w_{ij}^-}{\sum_i x_i w_{ij} - l_i w_{ij}^+ - h_i w_{ij}^-} R_j$	First

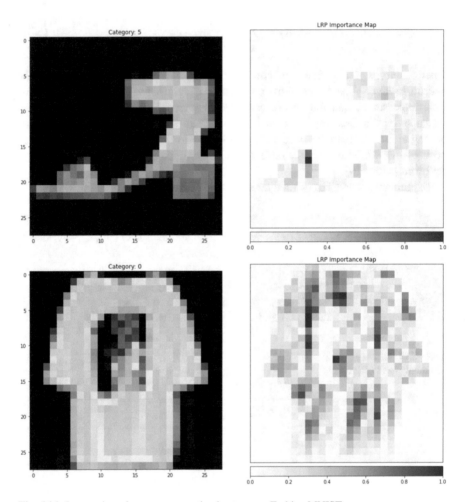

Fig. 6.16 Layer-wise relevance propagation heatmap on Fashion MNIST

Observations:

- Fig. 6.16 shows the category prediction of a PyTorch 3-layer CNN model trained on Fashion MNIST.
- The right side shows the layer-wise relevance propagation heatmap using a zero baseline and 50-step Riemann approximation.
- Note the quality of the visual explanations in comparison to DeepLIFT.

Table 6.18 Explainable
properties of layer-wise
relevance propagation

Properties	Values
Local or global	Local
Linear or non-linear	Non-linear
Monotonic or non-monotonic	Non-monotonic
Feature interactions captured	Yes
Model complexity	Medium

6.5.10 Excitation Backpropagation

Excitation backpropagation (EBP) [Zha+16] is an explanation method that aims
to visualize neuron activations by applying a winner-take-all (WTA) approach to
backpropagating through excitatory connections between neurons for classification
tasks. It backpropagates only positive weights while keeping gradients normalized.

For a neuron a_i in the i-th layer, the conditional winning probability $P(a_j|a_i)$ of
each neuron a_j in the preceding layer connected to it is

$$P(a_j|a_i) = \begin{cases} Z_i \hat{a}_j w_{ji} & w_{ji} \geq 0 \\ 0 & otherwise \end{cases} \tag{6.38}$$

where $\hat{a}_j$ is the input neuron's response and the normalization factor Z_i is given by

$$Z_i = \begin{cases} 0 & \sum_{j:w_{ji} \geq 0} \hat{a}_j w_{ji} = 0 \\ \dfrac{1}{\sum_{j:w_{ji} \geq 0} \hat{a}_j w_{ji}} & otherwise \end{cases} \tag{6.39}$$

In the winner-take-all approach, if a_i is a winning neuron, the next winning
neuron will be sampled based on $P(a_j|a_i)$. The weight w_{ji} reflects the top-down
feature expectation and $\hat{a}_j$ captures the bottom-up feature strength. Applying this
recursively allows EBP to compute marginal winning probability maps which can
serve as soft attention maps.

> Interpretation of Excitation Backpropagation: EBP learns soft attention maps
> by applying a probabilistic winner-take-all process to backpropagate activa-
> tions top-down through the network. It can also learn contrastive attention
> maps that improve discriminative ability. EBP is restricted to neural classifi-
> cation tasks.

In practice, EBP is often used to propagate a pair of contrastive top-down
signals by backpropagating both a positive and a negative neural activations top-
down through the network. As marginal winning probability maps are linear

functions of the top-down signal, the sum of these two activations can be computed simultaneously during a single backward pass. The resulting contrastive marginal winning probability map can amplify discriminative excitations.

Explainable properties of Excitation Backpropagation are shown in Table 6.19.

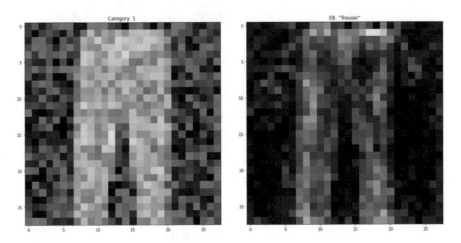

Fig. 6.17 Excitation backpropagation heatmap on Fashion MNIST

Observations:

- Fig. 6.17 shows the input image to a 3-layer MLP model trained on Fashion MNIST.
- The right side shows the excitation backpropagation soft attention map.

Table 6.19 Explainable properties of excitation backpropagation

Properties	Values
Local or global	Local
Linear or non-linear	Non-linear
Monotonic or non-monotonic	Non-monotonic
Feature interactions captured	Yes
Model complexity	Medium

6.5.11 CAM

Class activation maps (CAM) [Zho+15a] is an explanation method applicable to specific CNN architectures. It has been shown that CNNs can be used for object localization if the max pooling operation is replaced with global average pooling [Zho+15b]. By adding a global average pooling operation between the last convolutional layer and the output layer of a CNN, the discriminative image regions associated with a prediction for a particular class can be visualized as a class activation map.

Let $f_k(x, y)$ represent the activation of the k-th neuron in the last convolutional layer of a CNN at spatial location (x, y). Then the output P_c for class c is given by

$$P_c = \frac{\exp(S_c)}{\sum_c \exp(S_c)} \tag{6.40}$$

$$S_c = \sum_k w_k^c F_k \tag{6.41}$$

$$F_k = \sum_{x,y} f_k(x, y) \tag{6.42}$$

where w_k^c is the weight corresponding to class c for the k-th neuron, S_c is the input to the softmax neuron for class c, and F_k is the global average pooled output at the k-th neuron. The weights w_k^c can be interpreted as a measure of the importance of F_k for class c.

The class activation map $M_c(x, y)$ is defined as

$$M_c(x, y) = \sum_k w_k^c f_k(x, y) \tag{6.43}$$

Note that the class c softmax input can be written as

$$S_c = \sum_{x,y} \sum_k w_k^c f_k(x, y) = \sum_{x,y} M_c(x, y) \tag{6.44}$$

and the class activation map can be interpreted as a measure of importance of the activation at spatial location (x, y) for class c prediction. By up-sampling the class activation map to the size of the input image and applying thresholding, a heatmap is generated that identifies the regions of the input image most relevant to class c.

While CAM is computationally efficient as it requires a forward pass and a partial backward pass. Unfortunately, it is restricted to a set of specific CNN architectures that exclude fully connected layers.

Explainable properties of CAM are shown in Table 6.20.

Interpretation of Class Activation Maps: CAMs are useful for a specific set of CNN models by using global max pooling to visualize the regions of an input image most relevant to a class prediction. It exploits the spatial information that is preserved through convolutional layers.

Table 6.20 Explainable properties of CAM

Properties	Values
Local or global	Local
Linear or non-linear	Non-linear
Monotonic or non-monotonic	Non-monotonic
Feature interactions captured	Yes
Model complexity	Medium

6.5.12 Gradient-Weighted CAM

Gradient-weighed CAM (GradCAM) [Sel+19] is a generalization of CAM to allow for more flexible CNN architectures. Instead of relying on a global average pooling after the last convolutional layer, it allows for any architecture as long as layers are differentiable. GradCAM assigns importance values to each neuron by utilizing the gradient information that flows into the last convolutional layer of the CNN.

Let y^c be the score for class c, and let A_{xy}^k be the feature map activations of a convolutional layer neuron at location (x, y). GradCAM calculates the neuron importance weights α_k^c by global average pooling the gradients:

$$\alpha_k^c = \frac{1}{Z} \sum_{x,y} \underbrace{\frac{\partial y^c}{\partial A_{xy}^k}}_{\text{gradient}} \tag{6.45}$$

where Z is a proportionality constant that can be disregarded since it is normalized out during visualization. With these alpha weights, a GradCAM localization heatmap L_c for class c is calculated by

$$L_c = ReLU \left(\sum_k \alpha_k^c A^k \right) \tag{6.46}$$

The ReLU operation is to ensure only positive importance values are emphasized. In effect, GradCAM takes the weighted sum of the feature map activations of the convolutional layer to generate gradient-weighted class activation maps. These class activation maps are up-sampled to the size of the input image to generate heatmaps of importance values.

> Interpretation of Gradient-Weighted Class Activation Maps: GradCAM is a generalization of CAM to a broader range of CNN architectures. It can efficiently generate localization heatmap explanations for specific instances.

Like CAM, GradCAM is computationally efficient and requires a single forward pass and a partial backward pass. Unlike CAM, it is applicable to a much broader range of CNN-based architectures.

Explainable properties of GradCAM are shown in Table 6.21.

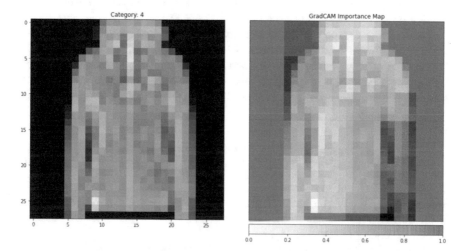

Fig. 6.18 Up-sampled localization heatmap on Fashion MNIST

Observations:

- Fig. 6.18 shows the input image to a PyTorch 3-layer CNN model trained on Fashion MNIST.
- The right side shows the up-sampled GradCAM heatmap for the conv3 layer (the last convolutional layer)
- GradCAM identifies the sleeve areas as important features to the prediction of category 4 (Coat).

Table 6.21 Explainable
properties of GradCAM

Properties	Values
Local or global	Local
Linear or non-linear	Non-linear
Monotonic or non-monotonic	Non-monotonic
Feature interactions captured	Yes
Model complexity	Medium

6.5.13 Testing with Concept Activation Vectors

One of the challenges of neural explanation methods is that they may not generate human-interpretable explanations. Feature importance heatmaps may identify regions of the input instance that influence output prediction, but these do not correspond to human-relatable concepts. Furthermore, hidden layer activations are seldom comprehensible. Concept activation vectors (CAVs) [Kim+18] map data and latent representations to human-interpretable concepts.

Let E_m represent the vector space of basis vectors e_m that span the input features and neural activations and E_h represent the vector space of human-interpretable concepts. A concept activation vector is a mapping from E_m to E_h and is learned by training a binary linear classifier f on the layer activations with a set of hand-selected positive examples that contain the concept, as well as a set of random negative instances:

$$f(x) = \begin{cases} 1 & w^T x + b \geq threshold \\ 0 & w^T x + b < threshold \end{cases} \tag{6.47}$$

where w and b are the weights and bias of the binary linear classifier. The CAV v_j^c is the normal vector to the learned hyperplane decision boundary in the direction toward the concept in the j-th layer, $v_j^c = w$.

Testing with Concept Activation Vectors (TCAV) is an explanation method that can quantify the class sensitivity of a trained neural network with respect to a concept in a neural network layer. Let the scoring function $S_{k,j}^c$ be defined as the sensitivity of the neural activation at the j-th layer to class k for a given instance x:

$$S_{k,j}^c = \nabla h_{k,j}(f_j(x)) v_j^c \tag{6.48}$$

where $f_j(x)$ maps the input x to the activation vector of layer j and $h_{k,j}$ maps the activation vector of layer j to the output activation (logit) of class k. The directional derivative is taken toward the concept activation vector v_j^c for layer j. This scalar represents the influence of concept c in influencing the model to predict x as class k. A positive value encourages while a negative value discourages the model toward class k.

TCAV measures the class sensitivity across inputs of an entire class at layer j by computing

$$TCAV_{k,j}^c = \frac{\left| x \in X_k : S_{k,j}^c > 0 \right|}{|X_k|} \tag{6.49}$$

where X_k is the set of input instances labeled as class k. This represents the fraction of instances with activation vectors at layer j positively influenced by concept c. It neither considers the magnitude of the influence nor negative influences.

TCAV has a distinct advantage in that concept activation vectors for user-defined concepts can be learned by providing examples from external datasets. Thus, it is possible to quantify the influence of semantic concepts that are much more human-comprehensible. However, not all concept activation vectors may be meaningful, since even a set of randomly selected instances can still produce a CAV. Furthermore, the CAVs learned for semantically opposing concepts may significantly overlap, resulting in less discriminative ability of the influence of related concepts.

> Interpretation of Testing with Concept Activation Vectors: concept activation vectors are effective representations of human-interpretable concepts in the activations of a neural network layer. These vectors can be used to test class sensitivity to particular concepts that generate meaningful and human-understandable explanations.

In practice, learned CAVs should be validated. One method is to retrain the CAV and calculate TCAV on multiple runs using different random negative instances. A meaningful CAV should result in consistent TCAV scores across these iterations, which can be evaluated using a t-test. Another way to validate CAVs is to visualize the patterns that activate each CAV by applying the activation maximization method. A further way to validate CAVs is to visualize the set of instances most and least similar to the CAV in terms of cosine distance.

Explainable properties of TCAV are shown in Table 6.22.

Observations:

- Fig. 6.19 shows the input image (category 9: ankle boot) and individual layer activations of a PyTorch 3-layer CNN model trained on Fashion MNIST.
- TCAV values were calculated using an SGD classifier for a "sneaker" concept trained on 100 positive samples with 100 random negative samples extracted from the test set.
- The TCAV values indicate that class 9 (ankle boot) is fairly sensitive to the concept "sneaker." From a human perspective, they share a semantic relationship as both are types of shoes.

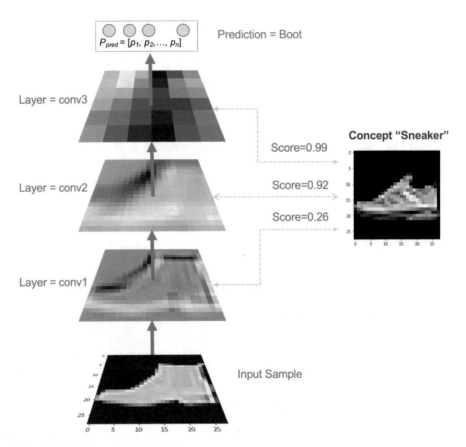

Fig. 6.19 TCAV on Fashion MNIST

Table 6.22 Explainable
properties of TCAV

Properties	Values
Local or global	Global
Linear or non-linear	Linear
Monotonic or non-monotonic	Monotonic
Feature interactions captured	Yes
Model complexity	High

References

[AJ18]   D. Alvarez-Melis, T.S. Jaakkola, *Towards robust interpretability with self-explaining neural networks* (2018). arXiv:1806.07538 [cs.LG]

[Bac+15a] S. Bach et al., On pixel-wise explanations for non-linear classifier decisions by layerwise relevance propagation. PLOS ONE **10**(7), 1–46 (2015). https://doi.org/10.1371/journal.pone.0130140.9

[Bin+16]   A. Binder et al., Layer-wise relevance propagation for deep neural network architectures, in *Information Science and Applications (ICISA) 2016*, ed. by K.J. Kim, N. Joukov, vol. 376. Lecture Notes in Electrical Engineering (Springer Singapore, Singapore, 2016), pp. 913–922. ISBN:978-981-10-0557-2

[Che+19]   C. Chen et al., *This looks like that: Deep learning for interpretable image recognition* (2019). arXiv:1806.10574 [cs.LG]

[CLL19]   H. Chen, S. Lundberg, S.-I. Lee, *Explaining models by propagating Shapley values of local components* (2019). arXiv:1911.11888 [cs.LG]

[Don+17]   Y. Dong et al., *Improving interpretability of deep neural networks with semantic information* (2017). arXiv:1703.04096 [cs.CV]

[ECB10]   D. Erhan, A. Courville, Y. Bengio, *Understanding Representations Learned in Deep Architectures*. Tech. rep. 1355. Université de Montréal/DIRO (Oct. 2010)

[FV17]   R.C. Fong, A. Vedaldi, Interpretable explanations of black boxes by meaningful perturbation, in *2017 IEEE International Conference on Computer Vision (ICCV)* (Oct. 2017). https://doi.org/10.1109/iccv.2017.371. http://dx.doi.org/10.1109/ICCV.2017.371

[Goy+20]   Y. Goyal et al., *Explaining classifiers with causal concept effect (CaCE)* (2020). arXiv:1907.07165 [cs.LG]

[HVD15]   G. Hinton, O. Vinyals, J. Dean, *Distilling the knowledge in a neural network* (2015). arXiv:1503.02531 [stat.ML]

[Iye+18]   R. Iyer et al., *Transparency and explanation in deep reinforcement learning neural networks* (2018). arXiv:1809.06061 [cs.LG]

[Kim+18]   B. Kim et al., Interpretability beyond feature attribution: Quantitative testing with concept activation vectors (TCAV), in *ICML*, ed. by J.G. Dy, A. Krause, vol. 80. Proceedings of Machine Learning Research (PMLR, 2018), pp. 2673–2682

[LBJ16]   T. Lei, R. Barzilay, T. Jaakkola, Rationalizing neural predictions. In *Proceedings of the 2016 Conference on Empirical Methods in Natural Language Processing* (Association for Computational Linguistics, Austin, Texas, 2016), pp. 107–117. https://doi.org/10.18653/v1/D16-1011. https://www.aclweb.org/anthology/D16-1011

[LMJ17]   J. Li, W. Monroe, D. Jurafsky, *Understanding neural networks through representation erasure* (2017). arXiv:1612.08220 [cs.CL]

[Li+17]   O. Li et al., *Deep learning for case-based reasoning through prototypes: a neural network that explains its predictions* (2017). arXiv:1710.04806 [cs.AI]

[LYW19]   H. Liu, Q. Yin, W.Y. Wang, *Towards explainable NLP: A generative explanation framework for text classification* (2019). arXiv:1811.00196 [cs.CL]

[Mon+19]   G. Montavon et al., Layer-wise relevance propagation: An overview. *Explainable AI* (2019)

[PDS18a]   V. Petsiuk, A. Das, K. Saenko, *RISE: Randomized input sampling for explanation of black-box models* (2018). arXiv:1806.07421 [cs.CV]

[Sel+19]   R.R. Selvaraju et al., Grad-CAM: Visual explanations from deep networks via gradient-based localization. Int. J. Comput. Vis. **128**(2), 336–359 (2019). ISSN: 1573-1405. http://doi.org/10.1007/s11263-019-01228-7

[SS19]   S. Serrano, N.A. Smith, *Is attention interpretable?* (2019). arXiv:1906.03731 [cs.CL]

[SGK19]   A. Shrikumar, P. Greenside, A. Kundaje, *Learning important features through propagating activation differences* (2019). arXiv:1704.02685 [cs.CV]

[SVZ14a] K. Simonyan, A. Vedaldi, A. Zisserman, *Deep inside convolutional networks: Visualising image classification models and saliency maps* (2014). arXiv:1312.6034 [cs.CV]

[Spr+15b] J.T. Springenberg et al., *Striving for simplicity: The all convolutional net* (2015). arXiv:1412.6806 [cs.LG]

[STY17]   M. Sundararajan, A. Taly, Q. Yan, *Axiomatic attribution for deep networks* (2017). arXiv:1703.01365 [cs.LG]

[Tur95]   A.M. Turing, *Computers & amp; thought* (MIT Press, 1995), pp. 11–35. Chap. Computing Machinery and Intelligence

[ZTF11]   M.D. Zeiler, G.W. Taylor, R. Fergus, Adaptive deconvolutional networks for mid and high level feature learning, in *2011 International Conference on Computer Vision* (2011), pp. 2018–2025. https://doi.org/10.1109/ICCV.2011.6126474

[ZF13]   M.D. Zeiler, R. Fergus, *Visualizing and understanding convolutional networks* (2013). arXiv:1311.2901 [cs.CV]

[Zel+19] R. Zellers et al., From recognition to cognition: Visual commonsense reasoning, in *2019 IEEE/CVF Conference on Computer Vision and Pattern Recognition (CVPR)*, June 2019, pp. 6713–6724. https://doi.org/10.1109/CVPR.2019.00688

[Zha+16] J. Zhang et al., *Top-down neural attention by excitation backprop* (2016). arXiv:1608.00507 [cs.CV]

[Zho+15a] B. Zhou et al., *Learning deep features for discriminative localization* (2015). arXiv:1512.04150 [cs.CV]

[Zho+15b] B. Zhou et al., *Object detectors emerge in deep scene CNNs* (2015). arXiv:1412.6856 [cs.CV]

[Zin+17] L.M. Zintgraf et al., *Visualizing deep neural network decisions: Prediction difference analysis* (2017). arXiv:1702.04595 [cs.CV]

Chapter 7
Explainability in Time Series Forecasting, Natural Language Processing, and Computer Vision

Various domains such as computer vision, natural language processing, and time series analysis have extensively applied machine learning algorithms in recent years. This chapter will discuss the research and applications of the interpretable and explainable algorithms in this domain. We will start with a time series algorithm survey, starting from traditional interpretable statistical models to modern deep learning algorithms. Next, we discuss NLP applications and the role of interpretability. Finally, we cover computer vision and how explainability has been a focus of considerable research. We will present a case study in each domain where the reader can get practical and real-world insights.

7.1   Time Series Forecasting

Forecasting, based on historical data, is one of the most critical applications of time series. There is a particular class of much easier problems, such as predicting daily temperature based on the last few days, while specific issues such as predicting volatility in the foreign exchange rates may be trickier. Understanding the factors, how they impact the target, seasonality, trend, etc., contribute to forecasting model quality. Time series modeling and forecasting have many parallels to the general machine learning process of training and predicting out-of-sample. This section will discuss the similarity and highlight the differences for time series modeling, especially from the explainability standpoint.

Table 7.1 Tools and libraries for time series forecasting

Tools and libraries	Description
statsmodels https://github.com/statsmodels/statsmodels	Implementations for most of the statistical time series models like ARIMA, SARIMA, VAR, exponential smoothing, etc.
pysf https://github.com/alan-turing-institute/pysf	Supervised forecasting for time series
sktime https://github.com/alan-turing-institute/sktime	Support for time series algorithms and scikit-learn compatible tools to build, tune and validate time series models for classification, regression, and forecasting.
DeepSeries https://github.com/EvilPsyCHo/Deep-Time-Series-Prediction	Deep Learning Models for time series prediction
Prophet https://facebook.github.io/prophet/	Scalable time series forecasting package
N-Beats https://github.com/philipperemy/n-beats	Neural basis expansion analysis for interpretable time series forecasting

7.1.1   Tools and Libraries

The following are some of the known tools and libraries that can be used for time series forecasting from a statistical, machine learning, and deep learning perspective (Table 7.1).

7.1.2   Model Validation and Evaluation

In generic machine learning, we divide the dataset into training, validation, and testing for modeling, hyperparameter selection, and estimations, respectively. We follow a similar process in time series with a few modifications. We cannot take random samples from the dataset to create the training/validation/test subsets, which assumes each instance in the dataset is independent. There is a dependence between the instances in time series as compared to normal tabular data, and the temporal order has to be maintained when creating the splits. The splits also need to consider the seasonality, trend, and other factors evident from the EDA process. Generally, the two most prevalent techniques are single and multiple train-validation-test split that respect the observations' temporal order. In certain cases where the predictions become less accurate over time, it is common to re-train the model with new data as and when available for further predictions; this is called walk-forward validation [HA18a].

7.1.3 Model Metrics

Time series forecasting commonly focuses on predicting real values and has a direct mapping to regression problems. In this section, we will focus on methods for evaluating real-valued predictions in time series.

Given a length-H forecast horizon a length-T training data series history $[y_1, \ldots, y_T] \in \mathbb{R}^T$, the task is to predict the future values $\mathbf{y} \in \mathbb{R}^H = [y_{T+1}, \ldots, y_{T+H}]$. The predictions are given by $\hat{\mathbf{y}} \in \mathbb{R}^H = [y_{T+1}, \ldots, y_{T+H}]$

1. Forecast Error (Residual Forecast error) The forecast error is the difference between the expected value and the predicted value. It is given by

$$e_{T+h} = y_{t+h} - \hat{y}_{t+h} \tag{7.1}$$

The units of the forecast error are the same as the units of the prediction. It is important to note the difference between the residuals and the forecast errors. We calculate the residuals on the training set while forecast errors are calculated on the test set. Also, residuals are based on one-step forecasts, while forecast errors can involve multi-step forecasts.

2. Mean Forecast error (Forecast Bias) Mean forecast error (MFE), also known as the forecast bias, is the average of the forecast error values. It is given by

$$MFE = \frac{1}{H} \sum_{h=1}^{H} y_{t+h} - \hat{y}_{t+h} \tag{7.2}$$

A mean forecast error value other than zero suggests the model's tendency to over forecast (negative error) or under forecast (positive error). A forecast error with a value of zero or close to zero indicates an unbiased model.

3. Mean Absolute Error The mean absolute error, or MAE, is the forecast error values' average, with absolute values used for all forecast predictions. When comparing forecast models applied to a single or multiple time series with the same units, the MAE is a popular metric as it is easy to understand and compute.

$$MFE = \frac{1}{H} \sum_{h=1}^{H} |y_{t+h} - \hat{y}_{t+h}| \tag{7.3}$$

4. Mean Squared Error (MSE) and Root Mean Squared Error (RMSE) The mean squared error, or MSE, is the average of the squared forecast error values. Squaring the forecast error values forces them to be positive. The side effect is that very large or outlier forecast errors get squared, which results in a larger mean squared error. Root mean squared error (RMSE) is measured by taking the square root of MSE.

$$MSE = \frac{1}{H} \sum_{h=1}^{H} (y_{t+h} - \hat{y}_{t+h})^2 \tag{7.4}$$

$$RMSE = \sqrt{\frac{1}{H} \sum_{h=1}^{H} (y_{t+h} - \hat{y}_{t+h})^2} \tag{7.5}$$

5. Mean Absolute Percentage Error and Scaled Mean Absolute Percentage Error
   Mean Absolute Percentage Error (MAPE) is given by

$$MAPE = \frac{100}{H} \sum_{h=1}^{H} \frac{|y_{t+h} - \hat{y}_{t+h}|}{|y_{t+h}|} \tag{7.6}$$

It is unit-free and can be used to compare across datasets. One of the limitations
of MAPE is that when the y_t is closer to 0 it causes a higher penalty on the
negative errors than positives. Symmetric MAPE (sMAPE) overcomes these
limitations of MAPE, but Hyndman et al. recommend against using sMAPE.

$$sMAPE = \frac{200}{H} \sum_{h=1}^{H} \frac{|y_{t+h} - \hat{y}_{t+h}|}{|y_{t+h}| + |\hat{y}_{t+h}|} \tag{7.7}$$

6. Mean Absolute Scaled Error Scaling the metric with the data is another common
   technique to make it independent of the units and recommended by Hyndman et
   al. as an alternative to using percentage errors when comparing forecast accuracy.
   For non-seasonal the Mean Absolute Scaled Error (MASE) is given by

$$MASE = \frac{1}{H} \sum_{h=1}^{H} \frac{|y_{t+h} - \hat{y}_{t+h}|}{\frac{1}{T-1} \sum_{t=2}^{T} |y_t - y_{t-1}|} \tag{7.8}$$

and seasonal metric scales by the average error of the naive predictor that simply
copies the observation measured m periods in the past, to account for seasonality.

$$MASE(seasonal) = \frac{1}{H} \sum_{h=1}^{H} \frac{|y_{t+h} - \hat{y}_{t+h}|}{\frac{1}{T-m} \sum_{t=m+1}^{T} |y_t - y_{t-1}|} \tag{7.9}$$

7.1.4 Statistical Time Series Models

7.1.4.1 ARIMA Models

AutoRegressive Integrated Moving Average (ARIMA) usually refers to a class of statistical models for time series forecast and analysis. Box and Jenkins formulated a detailed process for identifying, estimating, and reviewing models for a time series dataset [Box+15]. ARIMA, the acronym, captures different components of time series modeling and is represented in literature as $ARIMA(p, d, q)$. The components are:

- Autoregression(AR): The model's ability to capture the dependent relationship between an observation and a fixed number of lagged observations of itself. The parameter p represents the lag or number of the observations included.

$$y_t = \phi_0 + \phi_1 y_{t-1} + \cdots + \phi_p y_{t-p} \tag{7.10}$$

  The equation captures linear relationship between current value regressed with p lagged values and the coefficients ϕ are regression coefficients.
- Integrated(I): The ability to make the model stationary by differencing the raw observations from observations at the previous time steps. The parameter d represents the degree of first differencing. The backward shift operator B has an effect of shifting data back one period. Thus, applying B shifts the data back as given by

$$B y_t = y_{t-1} \tag{7.11}$$

  The first difference or lag can be defined as

$$y_t' = y_t - y_{t-1} = y_t - B y_t = (1 - B) y_t \tag{7.12}$$

  Thus the d-th order difference can be written as $(1 - B)^d y_t$
- Moving Average(MA): The ability to use past forecast errors from moving average in a regression-like model. The parameter q represents the moving average window's size, also called the order of moving average.

$$y_t = c + \varepsilon_t + \theta_1 \varepsilon_{t-1} + \cdots + \theta_q \varepsilon_{t-q} \tag{7.13}$$

  The equation captures the linear relationship between the forecast variable and the past forecast errors given by ε_t.

If we combine all the components,

$$y_t' = c + \phi_1 y_{t-1}' + \cdots + \phi_p y_{t-p}' + \varepsilon_t + \theta_1 \varepsilon_{t-1} + \cdots + \theta_q \varepsilon_{t-q} \tag{7.14}$$

Table 7.2 ARIMA models

Models	Arima(p,d,q)
White noise	Arima(0,0,0)
Random walk	Arima(0,1,0) no constant
Random walk with drift	Arima(0,1,0) with constant
Autoregression	Arima(p,0,0)
Moving Average	Arima(0,0,q)

The equation can be rewritten using the backshift operator as

$$\underbrace{(1 - \phi_1 B - \cdots - \phi_p B^p)}_{AR(p)} \underbrace{(1 - B)^d y_t}_{d\ differences} = \underbrace{c + (1 + \theta_1 B + \cdots + \theta_q B^q)\varepsilon_t}_{MA(q)}$$

(7.15)

Many non-seasonal models can be just represented as special cases of $ARIMA(p, d, q)$ as given in Table 7.2.

A seasonal ARIMA model (SARIMA) is formed by including additional seasonal terms in the ARIMA model. It is represented as:

$$ARIMA \underbrace{(p, d, q)}_{non-seasonal} \underbrace{(P, D, Q)_m}_{seasonal}$$

(7.16)

where (P, D, Q) capture the seasonal part and m is the number of observations per year.

For a given model parameters $Arima(p, d, q)(P, D_Q)_m$, maximum likelihood estimation (MLE) is used to estimate the values for$(c, \phi_1, \ldots, \phi_p, \theta_1, \ldots, \theta_q)$. Step-wise search is usually performed to find the best parameters p, d, q, P, D, Q, m and minimizing Akaike's Information Criterion (AIC) is used to select the best parameters.

AIC for ARIMA is given as

$$AIC = -log(L) + 2(p + q + k + 1)$$

(7.17)

We use the Mauna Loa CO2 dataset for ARIMA analysis. The data is resampled as monthly and 1958–2009, i.e., 610 months, are used for training. 2009–2017, i.e., 98 months, are used for testing and forecasting. The package pmdarima is used for performing the parameter selection and the output is given in Fig. 7.1.

```
Performing stepwise search to minimize aic
 ARIMA(1,1,1)(0,1,1)[12]                     : AIC=373.782, Time=0.77 sec
 ARIMA(0,1,0)(0,1,0)[12]                     : AIC=725.657, Time=0.04 sec
 ARIMA(1,1,0)(1,1,0)[12]                     : AIC=546.596, Time=0.25 sec
 ARIMA(0,1,1)(0,1,1)[12]                     : AIC=385.612, Time=0.67 sec
 ARIMA(1,1,1)(0,1,0)[12]                     : AIC=664.737, Time=0.12 sec
 ARIMA(1,1,1)(1,1,1)[12]                     : AIC=375.773, Time=1.34 sec
 ARIMA(1,1,1)(0,1,2)[12]                     : AIC=375.772, Time=3.00 sec
 ARIMA(1,1,1)(1,1,0)[12]                     : AIC=523.661, Time=0.45 sec
 ARIMA(1,1,1)(1,1,2)[12]                     : AIC=377.741, Time=4.54 sec
 ARIMA(1,1,0)(0,1,1)[12]                     : AIC=399.951, Time=0.61 sec
 ARIMA(2,1,1)(0,1,1)[12]                     : AIC=375.757, Time=2.86 sec
 ARIMA(1,1,2)(0,1,1)[12]                     : AIC=375.762, Time=1.29 sec
 ARIMA(0,1,0)(0,1,1)[12]                     : AIC=inf, Time=0.42 sec
 ARIMA(0,1,2)(0,1,1)[12]                     : AIC=376.202, Time=0.77 sec
 ARIMA(2,1,0)(0,1,1)[12]                     : AIC=389.050, Time=0.79 sec
 ARIMA(2,1,2)(0,1,1)[12]                     : AIC=377.579, Time=3.69 sec
 ARIMA(1,1,1)(0,1,1)[12] intercept           : AIC=inf, Time=2.97 sec

Best model:  ARIMA(1,1,1)(0,1,1)[12]
Total fit time: 24.598 seconds
```

Fig. 7.1 Auto ARIMA selecting parameters for minimal AIC

Interpretation of ARIMA Models:

1. Log-Likelihood, AIC, BIC, and HQIC are different measures that assess the quality of the models. Lower the value, better is the model.
2. Different model coefficients like autoregression ($ar.L.p$), moving average ($ma.L.q$) and seasonal ($mas.S.L.m$) along with standard errors, z-value, p-value and confidence intervals (25–75%) the actual regression value and the significance. The interpretation is: increasing the coefficient by one unit changes the estimated outcome by its weight value.
3. The Ljung-Box test output is a diagnosis test for autocorrelation. The null hypothesis being the data is independently distributed, and there is no autocorrelation, while the alternate hypothesis is that data is not independently distributed and there is serial correlation. The p-value (Prob(Q)) less than $p < 0.05$ is generally for significance testing and rejecting the null hypothesis
4. The Jarque-Bera is a diagnosis test and is a type of Lagrange multiplier test for normality with the null hypothesis being the data is normally distributed and the alternate hypothesis not normally distributed. The p-value (Prob(JB)) less than $p < 0.05$ is generally for significance testing and rejecting the null hypothesis.
5. The heteroskedasticity is a test for heteroskedasticity of standardized residuals, null hypothesis is there being no heteroskedasticity and the

(continued)

alternate being presence of heteroskedasticity. The p-value (Prob(H)) less than $p < 0.05$ is generally for significance testing and rejecting the null hypothesis.
6. Diagnosis plots also help interpret the ARIMA models. Residual plot is a plot of standardized residuals against time. If the plot shows a scatter around the zero levels, it indicates that the specified model adequately captures variations. The KDE curve should be very similar to the normal distribution, indicating the errors are normally distributed. A normal Q-Q plot of the residuals is a graphical check for normal errors. If the Q-Q plot resembles a straight line, then the assumption that the errors are normally distributed is valid. Correlogram captures the correlation.

Observations:

- Figure 7.1 shows how the auto ARIMA process finds the best parameters $Arima(1, 1, 1)(0, 1, 1)[12]$ for the dataset.
- Figure 7.2 shows all coefficients being significant and having values in 25–75% confidence intervals.
- The p-value is 0.67 and is >0.05 for the Ljung-Box test, which means we cannot reject the null hypothesis. It means that the residuals are independent which further indicates that the model provides an adequate fit to the data (Fig. 7.3).
- The p-value is 0 and is <0.05 for the Jarque-Bera test for normality, indicating we can reject the null hypothesis that the data is normally distributed.
- The p-value is 0.02 and is <0.05 for the heteroskedasticity, indicating we can reject the null hypothesis and that the variance of the residuals is not constant.
- Figure 7.4 residual plots show residuals are scattered around zero level and have no pattern, indicating the model's ability to regress well on the data.
- Figure 7.4 normal Q-Q plot shows the points not on the straight line. This indicates not normal distribution or having a fat tail.

7.1.4.2   Exponential Smoothing Models

Exponential smoothing is one of the most successful forecasting methods and has driven many applications [Bro59, Hol04, Win60]. In exponential smoothing methods, forecasts are weighted averages of past observations, with the weights

SARIMAX Results

Dep. Variable:			y	No. Observations:		610
Model:	SARIMAX(1, 1, 1)x(0, 1, 1, 12)			Log Likelihood		-182.891
Date:	Wed, 17 Feb 2021			AIC		373.782
Time:	12:58:46			BIC		391.349
Sample:	0			HQIC		380.622
	- 610					
Covariance Type:	opg					

	coef	std err	z	P>|z|	[0.025	0.975]
ar.L1	0.3802	0.077	4.934	0.000	0.229	0.531
ma.L1	-0.6887	0.064	-10.812	0.000	-0.814	-0.564
ma.S.L12	-0.8750	0.025	-35.374	0.000	-0.923	-0.827
sigma2	0.1049	0.004	24.567	0.000	0.097	0.113

Ljung-Box (Q):	35.64	Jarque-Bera (JB):	219.09
Prob(Q):	0.67	Prob(JB):	0.00
Heteroskedasticity (H):	0.72	Skew:	0.40
Prob(H) (two-sided):	0.02	Kurtosis:	5.86

Fig. 7.2 ARIMA model on CO2 data

Fig. 7.3 ARIMA model on CO2 data

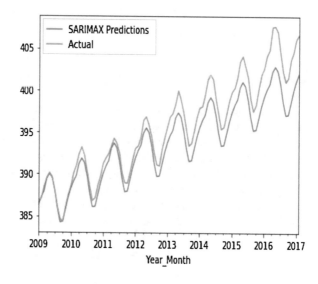

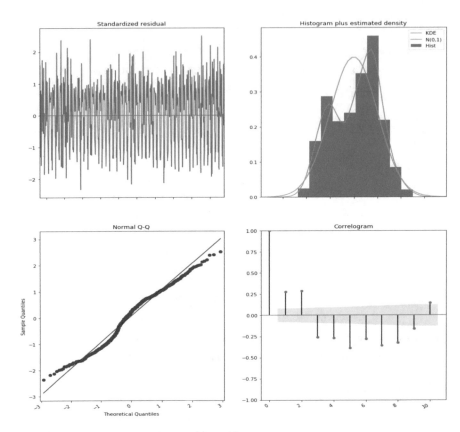

Fig. 7.4 Diagnostics plots for the ARIMA model

decaying exponentially as the observations get more distant from the forecasting period.

Holt-Winters' method is a popular exponential smoothing technique that captures different aspects of past observations (level), trend, and seasonality in the time series. This method has two variations based on how the seasonal components get applied. The additive method is used when the seasonal variations are almost constant, while the multiplicative method is preferred when the seasonal variations change proportionally to the series' level.

The additive method of forecasting y_t is given in terms of the level equation l_t, the trend equation b_t and the seasonality equation s_t.

$$\hat{y}_{t+h|t} = l_t + hb_t + s_{t+h-m(k+1)} \tag{7.18}$$

where m is a seasonal component and $k = \frac{h-1}{m}$. The level equation given below is linear weighted average between the seasonally adjusted observation $(y_t - s_{t-m})$ and the non-seasonal part $(l_{t-1} + b_{t-1})$ using the parameter α

$$l_t = \alpha(y_t - s_{t-m}) + (1 - \alpha)(l_{t-1} + b_{t-1}) \tag{7.19}$$

b_t denotes an estimate of the trend (slope) of the series at time t and is measured a weighted average between $(l_t - l_{t-1})$ and b_{t-1}. The reason β^* which has relationship $\beta = \alpha\beta^*$ is used for simplification.

$$b_t = \beta^*(l_t - l_{t-1}) + (1 - \beta^*)b_{t-1} \tag{7.20}$$

The seasonal component s_t is expressed in terms of the weighting between level l_{t-1} and the trend b_{t-1} and the previous seasonal components s_{t-m}

$$s_t = \gamma(y_t - l_{t-1} - b_{t-1}) + (1 - \gamma)s_{t-m} \tag{7.21}$$

The Holt-Winters' multiplicative method expresses the seasonal component in relative terms by dividing through by the seasonal component as given below:

$$\hat{y}_{t+h|t} = (l_t + hb_t)s_{t+h-m(k+1)} \tag{7.22}$$

$$l_t = \alpha\frac{y_t}{s_{t-m}}s_{t-m} + (1 - \alpha)(l_{t-1} + b_{t-1}) \tag{7.23}$$

$$b_t = \beta^*(l_t - l_{t-1}) + (1 - \beta^*)b_{t-1} \tag{7.24}$$

$$s_t = \gamma\frac{y_t}{(l_{t-1} - b_{t-1})} + (1 - \gamma)s_{t-m} \tag{7.25}$$

We can write many known exponential smoothing methods as below by analyzing variations in the combinations of trend and seasonal components (Table 7.3). If the time series shows different variations at different levels of the series, then transformations such as $w_t = \log(y_t)$, $w_t = \sqrt{y_t}$, etc. is used. Box-Cox transformation which represents a family of transformations based on the value of λ given below is a very known technique to apply with exponential smoothing.

Table 7.3 Variations and methods corresponding to each exponential smoothing method

Trends	Seasonal		
	None (N)	Additive (A)	Multiplicative (M)
None (N)	(N,N) Simple exponential smoothing	(N, A)	(N, M)
Additive (A)	(A, N) Holt's linear method	(A, A) Additive Holt-Winters' method	(A, M) Multiplicative Holt-Winters method
Additive damped (A_d)	(A_d, N) Additive damped method	(A_d, N)	(A_d, M) Holt-Winters' damped method

Explainability in Time Series Forecasting, Natural Language Processing, and...

$$w_t = \begin{cases} log(y_t) & \text{if } \lambda = 0 \\ (y_t^\lambda - 1) & \lambda \neq 0 \end{cases} \qquad (7.26)$$

Interpretation of Holt-Winters' Model: The values of α, β, and γ which should be between (0, 1) can be used to interpret the impact of level, trend, and seasonality. The value of α indicates how much the estimate of the level at the current time is based upon both recent observations and observations in the more distant past; larger the value, larger is the dependency. The value of β indicates the estimate of the slope b of the trend component, larger the value, faster is the rate of trend. The value of γ indicates how the estimate of the seasonal component at the current time point based upon very recent observations, higher the estimate, larger is the dependency.

Observations:

- We model Holt-Winters's on the training dataset using "Additive" for the trend part and "Multiplicative" for the seasonal part. We model one with the Box-Cox transformation and one without to show the impact.
- Figure 7.5 shows the outputs for both models with and without Box-Cox transformation. With the Box-Cox transformation, the α and the β are relatively lower as compared to without the transformations showing how the transformation reduces the smoothing effect further. The γ value being higher in the Box-Cox transformation than without indicates how the seasonal dependencies get more pronounced with the transformation compared to without.
- Figure 7.6 shows forecasts made with two Holt-Winters' models, one with Box-Cox transformation and one without. Clearly, the one with Box-Cox shows it is better in estimating and tracks the actual very closely.
- Table 7.4 shows different metrics for each method discussed. Holt-Winters' exponential smoothing with Box-Cox transformation yields the lowest error across each metrics and confirming the forecasts plots.

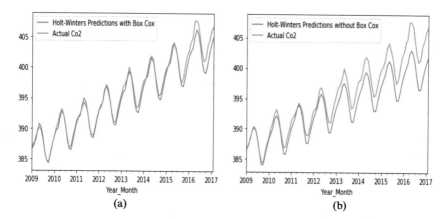

Fig. 7.5 Exponential smoothing and impact of Box-Cox transformation. (**a**) Forecasts with Box-Cox. (**b**) Forecasts without Box-Cox

ExponentialSmoothing Model Results

Dep. Variable:	endog	No. Observations:	610
Model:	ExponentialSmoothing	SSE	90.610
Optimized:	True	AIC	-1131.205
Trend:	Additive	BIC	-1060.590
Seasonal:	Multiplicative	AICC	-1130.048
Seasonal Periods:	12	Date:	Sun, 21 Feb 2021
Box-Cox:	True	Time:	14:28:09
Box-Cox Coeff.:	-2.26216		

	coeff	code	optimized
smoothing_level	0.3684211	alpha	True
smoothing_slope	0.0526316	beta	True
smoothing_seasonal	0.3684211	gamma	True
initial_level	0.4420550	l.0	True
initial_slope	4.4939e-10	b.0	True
initial_seasons.0	0.9999996	s.0	True
initial_seasons.1	0.9999996	s.1	True
initial_seasons.2	0.9999996	s.2	True
initial_seasons.3	0.9999996	s.3	True
initial_seasons.4	0.9999996	s.4	True
initial_seasons.5	0.9999996	s.5	True
initial_seasons.6	0.9999996	s.6	True
initial_seasons.7	0.9999996	s.7	True
initial_seasons.8	0.9999996	s.8	True
initial_seasons.9	0.9999996	s.9	True
initial_seasons.10	0.9999996	s.10	True
initial_seasons.11	0.9999996	s.11	True

(a)

ExponentialSmoothing Model Results

Dep. Variable:	endog	No. Observations:	610
Model:	ExponentialSmoothing	SSE	61.375
Optimized:	True	AIC	-1368.843
Trend:	Additive	BIC	-1298.228
Seasonal:	Multiplicative	AICC	-1367.686
Seasonal Periods:	12	Date:	Sun, 21 Feb 2021
Box-Cox:	False	Time:	14:28:28
Box-Cox Coeff.:	None		

	coeff	code	optimized
smoothing_level	0.6878678	alpha	True
smoothing_slope	0.0085494	beta	True
smoothing_seasonal	0.000000	gamma	True
initial_level	346.02073	l.0	True
initial_slope	0.0935103	b.0	True
initial_seasons.0	0.9124443	s.0	True
initial_seasons.1	0.9155438	s.1	True
initial_seasons.2	0.9166923	s.2	True
initial_seasons.3	0.9148943	s.3	True
initial_seasons.4	0.9108599	s.4	True
initial_seasons.5	0.9053386	s.5	True
initial_seasons.6	0.9006143	s.6	True
initial_seasons.7	0.9002442	s.7	True
initial_seasons.8	0.9033775	s.8	True
initial_seasons.9	0.9064241	s.9	True
initial_seasons.10	0.9088445	s.10	True
initial_seasons.11	0.9106788	s.11	True

(b)

Fig. 7.6 Exponential smoothing and impact of Box-Cox transformation. (**a**) Forecasts with Box-Cox. (**b**) Forecasts without Box-Cox

Table 7.4 Various metrics
for three exponential
smoothing methods

Methods	MAPE	MAE	MFE	RMSE
SARIMAX	0.5	1.9	1.9	2.3
Holt-Winters' without Box-Cox	0.5	2.0	1.9	2.4
Holt-Winters' with Box-Cox	0.2	0.7	0.6	0.9

7.1.5 Prophet: Scalable and Interpretable Machine Learning Approach

Prophet is an open-source time series forecasting algorithm from Facebook. Prophet's objective is to provide a solution for forecasting "at scale" that combines configurable models with analyst-in-the-loop [TL18]. Many statistical time series models such as ARIMA and Exponential Smoothing rely on experts to interpret, analyze, and adjust the parameters to model the time series. One of the critical design considerations in Prophet is to have intuitive parameters for the end-users, which one can modify without knowing the details of the underlying algorithms.

We can map the working of Prophet to Generalized Additive Model (GAM) with time as a regressor as given below:

$$y(t) = g(t) + s(t) + h(t) + \epsilon(t) \qquad (7.27)$$

The function $g(t)$ is the trend function that models non-periodic changes, the function $s(t)$ models the seasonality or periodic changes, the function $h(t)$ captures the irregular holiday season and $\epsilon(t)$ is for idiosyncratic changes that are not accommodated by the model.

The trend function can be modeled as a non-linear or linear function. The simplified non-linear trend function is given by

$$g(t) = \frac{C}{1 + \exp\left(-k(t - m)\right)} \qquad (7.28)$$

where C is the carrying capacity similar to people with internet access, k is the growth rate and m is the offset parameter. This can be made more dynamic by making both carrying capacity and growth rate as a function of time as given in the paper. The linear trend model is a simple piecewise linear function with a constant rate of growth.

The seasonality and periodic effects with smoothing effects is modeled in a flexible way using Fourier series as given below:

$$s(t) = \sum_{n=1}^{N} \left(a_n \cos\left(\frac{2\pi nt}{P}\right) + b_n \sin\left(\frac{2\pi nt}{P}\right) \right) \qquad (7.29)$$

The $2N$ parameters are estimated by constructing a matrix of seasonality vectors $X(t)$ for each t and the seasonal component then becomes

$$s(t) = X(t)\beta \tag{7.30}$$

where $\beta \sim Normal(0, \sigma^2)$ for smooth priors.

Holidays can be customized per country/region by the users. For each holiday i, if D_i is the set of past and future dates around that, then an indicator function represents whether time t is during holiday i. Each holiday is assigned a parameter κ_i corresponding to a change in the forecast.

$$Z(t) = [\mathbf{1}(t \in D_1), \cdots, \mathbf{1}(t \in D_L)] \tag{7.31}$$

$$h(t) = Z(t)\kappa \tag{7.32}$$

Similar to seasonality $\kappa \sim Normal(0, \nu^2)$

Interpretation of Prophet Model: For each forecast instance, prophet outputs various parameters that contribute to the level, trend, and seasonality. The overall equation from interpretability standpoint is given as

$$\hat{y} = trend * (1 + multiplicative\_terms) + additive\_terms \tag{7.33}$$

where the *multiplicative_terms* and *additive_terms* both are combinations of seasonal and regressor factors.

Observations:

- Figure 7.7 shows trend part with upper and lower bounds, the predicted target as $\hat{y}$, and multiplicative terms for the last 10 months.
- Figure 7.8 shows the seasonal part with columns for sine and cosine elements for the last 10 months for the Eq. (7.29).

7.1.6 Deep Learning and Interpretable Time Series Forecasting

Though deep learning models have become standard architectures for many natural language processing, speech recognition, and computer vision tasks, they struggle

	ds	yhat	yhat_lower	yhat_upper	trend	trend_lower	trend_upper	multiplicative_terms	multiplicative_terms_lower	multiplicative_terms_upper
698	2016-05-01	403.625728	400.967119	406.599086	400.803513	398.435713	403.580116	0.007041	0.007041	0.007041
699	2016-06-01	402.568238	400.024632	405.604870	400.969600	398.566652	403.811579	0.003987	0.003987	0.003987
700	2016-07-01	401.115835	398.292316	404.127216	401.130329	398.692404	404.058291	-0.000036	-0.000036	-0.000036
701	2016-08-01	399.622737	397.036436	402.882871	401.296415	398.786924	404.301143	-0.004171	-0.004171	-0.004171
702	2016-09-01	398.593871	395.790786	401.705376	401.462502	398.898354	404.509753	-0.007145	-0.007145	-0.007145
703	2016-10-01	398.355276	395.716668	401.605872	401.623231	399.020187	404.709334	-0.008137	-0.008137	-0.008137
704	2016-11-01	399.003430	396.250692	402.245485	401.789317	399.139050	404.916435	-0.006934	-0.006934	-0.006934
705	2016-12-01	400.372103	397.541816	403.557305	401.950046	399.256401	405.121072	-0.003926	-0.003926	-0.003926
706	2017-01-01	402.215146	399.277091	405.508632	402.116133	399.377752	405.320969	0.000246	0.000246	0.000246
707	2017-02-01	404.032031	401.118367	407.359430	402.282219	399.499103	405.534106	0.004350	0.004350	0.004350

Fig. 7.7 Trend with multiplicative terms

Fig. 7.8 Trend with seasonal sine and cosine terms

yearly_delim_1	yearly_delim_2
0.763889	-0.645348
0.329408	-0.944188
-0.179767	-0.983709
-0.655156	-0.755493
-0.948362	-0.317191
-0.981306	0.192452
-0.746972	0.664855
-0.321270	0.946988
0.205104	0.978740
0.674444	0.738326

in the domain of time series forecasting [MSA18, Mak+82]. In the recent M4 competition, less than 10% approaches used "pure" machine learning or deep learning, and also the top-performing algorithms were hybrids of classical statistical time series algorithms and deep learning techniques [Mak+82].

Oreshkin et al.'s neural basis expansion analysis for interpretable time series forecasting (N-BEATS) is the first pure deep learning-based approach that outperforms well-established statistical methods and has interpretable outputs similar to the traditional statistical techniques that practitioners want [Rem20].

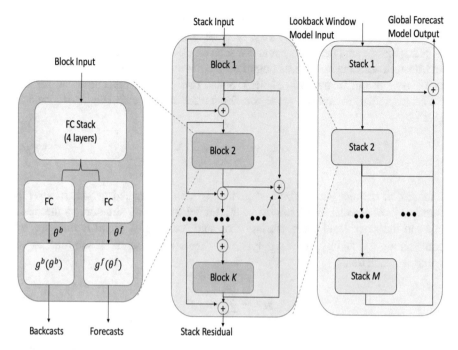

Fig. 7.9 N-BEATS neural architecture

Given a length-H forecast horizon, a length-T training data series history $[y_1, \ldots, y_T] \in \mathbb{R}^T$, the predictions given by $\hat{\mathbf{y}} \in \mathbb{R}^H = [y_{T+1}, \ldots, y_{T+H}]$, the forecast and the ability to lookback for a time window $t < T$ serves as the model input, given by $\mathbf{x} \in \mathbb{R}^T = [y_{T-t+1}, \ldots, y_T]$.

Figure 7.9 gives the detailed architecture of N-BEATS and we will start with the lth block. Each block accepts input $\mathbf{x}_l$ and outputs the forecast $\hat{\mathbf{y}}_l$ and the backcast $\hat{\mathbf{x}}_l$ (block's best estimate of its input $\mathbf{x}_l$). The block has four layers of fully connected networks and the final output is given by

$$\mathbf{h}_{l,4} = FC_{l,4}(FC_{l,3}(FC_{l,2}(FC_{l,1}(\mathbf{x}_l)))) \tag{7.34}$$

where each FC layer is a standard fully connected layer with ReLU non-linearity, for example, the first layer can be expanded as

$$FC_{l,1} = RELU(\mathbf{W}_{l,1}\mathbf{x}_l + b_{l,1}) \tag{7.35}$$

The basic building block then forks into two parts, the first being a fully connected, one for the forward expansion coefficients ($\theta_l^f = LINEAR_l^f(\mathbf{h}_{l,4})$) and another for the backward expansion coefficients ($\theta_l^b = LINEAR_l^b(\mathbf{h}_{l,4})$) using linear projections. The second part of the network maps expansion coefficients to the outputs via the basis functions, i.e., $\hat{\mathbf{y}}_l = g_l^f(\theta_l^f)$ and $\hat{\mathbf{x}}_l = g_l^b(\theta_l^b)$.

Next, the residual networks connect k such blocks. The standard residual network architecture adds the input to its output before passing the result to the next, proven to be very useful in training. N-BEATS has two residual branches, one for the backcast prediction and the other one for the forecast branch (Fig. 7.10), and is given by

$$\mathbf{x}_l = \mathbf{x}_{l-1} - \hat{\mathbf{x}}_{l-1}, \hat{\mathbf{y}} = \sum_l \hat{\mathbf{y}}_l \qquad (7.36)$$

Each block outputs partial forecast $\hat{\mathbf{y}}_l$ that is then aggregated to get the overall forecast $\hat{\mathbf{y}}$. Interpretability is added to the model by adding structure to the basis layer at the stack level for capturing trend and seasonality. Trend is modeled by constraining $g_{s,l}^b$ and $g_{s,l}^f$ to be a polynomial function of a degree p and function of window t such that:

$$\hat{\mathbf{y}}_{s,l} = \sum_{i=0}^{p} \theta_{s,l,i}^f t^i \qquad (7.37)$$

where $\mathbf{t} = [0, 1, 2, \ldots, H-2, H-1]^T / H$ is the time vector forecasting H steps. Similarly, seasonality is modeled as periodic function and is given by the Fourier series:

$$\hat{\mathbf{y}}_{s,l} = \sum_{i=0}^{\lfloor H/2-1 \rfloor} \theta_{s,l,i}^f \cos(2\pi i t) + \theta_{s,l,i}^f \sin(2\pi i t) \qquad (7.38)$$

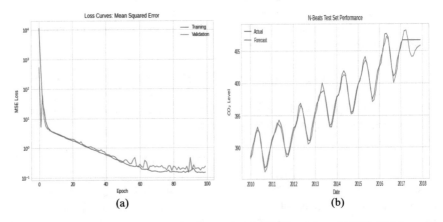

Fig. 7.10 N-BEATS on Mauna Loa CO2 dataset. (**a**) N-BEATS training. (**b**) N-BEATS forecasting

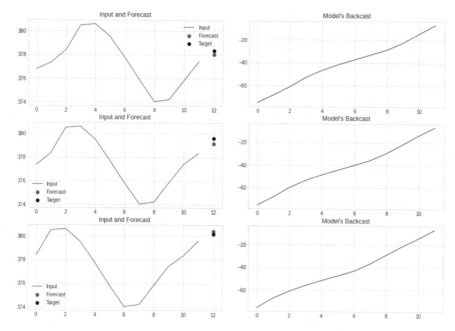

Fig. 7.11   N-BEATS interpretable model output

Figure 7.11 shows the output of interpretable model.

Observations:
Figure 7.11 highlights the smoothed trend over the input series, as seen in the first few entries in the validation set. Figure 7.10b shows that N-BEATS captures both trend and seasonality in the forecasting.

7.2   Natural Language Processing

Traditional Natural Language Processing (NLP) employed algorithms and methods that were more or less interpretable or explainable. Most techniques were explicit text features based on tokens or words, n-grams, part of speech, etc., for text classification, categorization, topic mining, question answering, entity recognition, summarization, etc. Employing standard feature relevancy techniques and interpretable algorithms such as logistic regression, decision trees, hidden Markov model, etc. made many tasks white box. In recent years, by achieving the state-of-the-art results, deep learning-based techniques have become standard implementation choices for most NLP tasks [KLW19b]. Since, by nature, most of the deep learning techniques are a black box, various techniques have evolved that focus on surrogate methods or some way of understanding the deep learning architectures in the context of NLP tasks. XAI has become a standard track in many NLP conferences attracting many research papers, libraries, and tools. Mapping XAI to NLP tasks and various related techniques have been presented in surveys and tutorials [Dan+20, AB18]. In this section, we will highlight some of the explanation methods and visualization techniques and their categorization.

7.2.1 Explainability, Operationalization, and Visualization Techniques

In most surveys and tutorials, XAI is presented through explainability, operations that enable explainability, and visualization techniques that assists in post-hoc or are intrinsic to the task as local or global explanations. We will combine these together under explainability techniques and give relevant operationalization and visualization methods that helps in enabling.

7.2.1.1 Feature Importance

Scoring text features (raw input texts or n-grams or latent features from layers of deep learning) in terms of relevancy to the NLP tasks is one of the fundamental explainable techniques [Vos+15, God+18]. One can apply various feature relevancy and scoring techniques discussed in Chap. 2 for traditional model pipelines. For deep learning techniques, understanding the latent feature representations, attention mechanisms and first-derivative saliency are the standard techniques used for feature importance [LWM19, Xie+17, BCB14].

In many sequence to sequence tasks such as machine translation, text summarization, question answering, etc., attention mechanism provides a way to score latent features over the decoded token. The weight distribution of these latent features that map directly or indirectly to input tokens for a given output helps identify essential features and visualize them from the input/output perspective.

Gradient-based explanations measure the importance of input towards output by computing the partial derivative of output with respect to input. The first-derivative saliency is effective as all modern deep learning frameworks provide auto-differentiation and can compute for any layers of deep learning architecture.

7.2.1.2 Surrogate Model

As discussed in the post-hoc model explainability chapter, surrogate techniques use interpretable models as a proxy to explain the original black-box model's behavior with the same inputs given to the black-box model and outputs from it.

Many techniques, such as LIME, SHAP, taxonomy through network embeddings, are broadly used as surrogate models [RSG16a, LL17a]. As discussed in Chap. 5, LIME provides an excellent local explanation and visualization by scoring the features that contribute to the model prediction. LIME can be used with any classifier models ranging from traditional Logistic regression to modern BERT-based classifiers. LIME explainer can help understand how the text features (raw tokens or mapped through embeddings) influence the outcome in text classification.

It has been found that by using different sampling around the same local instances, LIME explanation results can be entirely different and can therefore result in instability [Mol19]. LIME creates linear local models, and for many high-dimensional non-linear problems in NLP, this can be an issue.

As discussed in Chap. 5, SHAP uses cooperative game theory and Shapley values to model each feature's contribution to the final prediction. SHAP overcomes all of the disadvantages of LIME and is a more accepted explaining technique in practice, especially for text classification.

Liu et al. proposed interpreting network embeddings as a surrogate model technique for various machine learning problems, including NLP [Liu+18]. This surrogate model derives global explanation, unlike LIME and SHAP. The first task is to convert documents to vectors (e.g., tf-idf) and then use similarity techniques (e.g., cosine similarity) to create a network. Performing iterative clustering induces a taxonomy from the network embedding that is interpretable.

7.2.1.3 Example Driven

As the name suggests, these techniques take input and find examples from training data similar to the input and use the prediction and similarity for the explanation. This approach is very similar to the K-nearest neighbor approach and can be employed as the implementation technique to find similar examples from training as an explanation method [Dud76].

Croce et al. use the layer-wise relevance propagation technique in a kernel deep learning architecture to output prediction and examples that they call "landmarks" similar in semantic and linguistic properties as the input [CRB19]. The effectiveness of this technique is demonstrated in two NLP tasks, viz. question classification and semantic role labeling.

7.2.1.4 Provenance-Based

Provenance-based techniques work by demonstrating a series of reasoning steps that lead to the final prediction for a given instance. Many question-answering tasks use this technique along with knowledge bases. Provenance-based systems help diagnose when things go wrong and provide a hint as to how to reformulate the questions in question-answering tasks.

Abujabal et al. proposed Quint, a live system for explaining question answering using DBpedia. Quint shows how it understood the question by using entity linking and relation linking [Abu+17]. The entity linking extracts and disambiguates all the entities from the question and associates them with the knowledge base entities. Similarly, relation linking identifies and disambiguates relations from the question and matches it to knowledge base predicates. The derivation sequence shows how the entity, the relation mappings, and the SPARQL query lead to the final answer. The SPARQL queries are learned from the query templates based on structurally similar phrases.

Zhou et al. address the same explainability in question-answering tasks but on more complex questions with multiple relations than just one [ZHZ18]. The research builds an Interpretable Reasoning Network (IRN) that breaks down the knowledge base question answering (KBQA) into multi-hop reasoning steps, and in each step, the entities and their relations are disambiguated and predicted. An interesting aspect of this work is that it facilitates correction and disambiguation by humans.

7.2.1.5   Declarative Induction

Creating declarative specifications that are human-readable such as rules, trees, and programs to provide local or global explanations falls under a category known as declarative induction.

Jiang et al., in their research, showed how reasoning trees could connect and explore information from multiple sentences and documents to answer questions in the reading comprehension task [Jia+19]. The reasoning tree constructs many root-to-leaf paths representing the possible answer to the question. Based on the aggregation across different paths, a final answer is derived, and the path serves as an explanation for the choice and alternatives for diagnosing issues.

Entity resolution tasks (ER or record linking) in NLP have seen much success with various post-hoc explanation techniques, especially with declarative induction methods. ExplainER by Ebaid et al. uses Bayesian Rule List (BRL) to output declarative rules explaining the entity resolutions for global explanations [Eba+19].

Pezeshkpour et al. proposed declarative rules using adversarial modifications (e.g., removing facts) for link prediction tasks in knowledge graphs (KG). The method identifies KG nodes/facts that are most likely to influence the link prediction and then aggregates the extracted set of rules operating at the whole KG level for global explanations based on the frequent pattern of rules and the link [PTS19].

Sen et al. use linguistic rules to explain sentence classifications to the respective domains. The model is a neuro-symbolic, i.e., neural model to learn the rules model, where the rules not only perform classification but provide an explanation [Sen+19]. The rules are based on predicates, which depend on features resulting from syntactic and semantic parsing of the input sentence. The system provides flexibility to use a domain-specific dictionary, generalize better, and allows experts to interact and provide inputs.

7.2.2   *Explanation Quality Evaluation*

Evaluating the quality of metrics for NLP model explanations is similar to general XAI metrics and is an evolving field. We can broadly classify the techniques into the following two categories.

7.2.2.1   Comparison to the Ground Truth

The use of specific metrics depends on the particular NLP task but gives the advantage of automation. Carton et al. use deep adversarial networks for extracting explanations in classification and employ precision, recall, and F1 scores as quality metrics for the explanations [CMR18]. Rajani et al. use BLEU scores as metrics in their Commonsense Auto-Generated Explanation (CAGE) for CommonsenseQA task [Raj+19]. Quantitative metrics have an advantage over subjective qualifications, but they suffer from a disadvantage when there are alternate explanations or when the ground truth quality itself is in question. A couple of strategies such as—(a) Having multiple annotators to evaluate the explanation and using inter-annotator agreement for matching the ground truth. (b) Evaluating the explanations at different granularities are two different ways to overcome the disadvantages.

7.2.2.2   Human Evaluation

One natural but manual evaluation process is to use one or many experts to appraise the explanation. This technique avoids the assumption of the availability of only one reasonable explanation for the ground truth. This technique also needs to take into account inter-annotator disagreements and variances across them. Based on how many humans evaluate (single vs. multiple) and choice of single approach or multiple combined explanations, there has been much exciting research in the area [Mul+18, SPR19, Don+19].

7.2.3   *Tools and Libraries*

Table 7.5 provides details of the libraries used for NLP model explanations in the chapter.

7.2.4   *Case Study*

We will explore different tools and libraries to understand different classifier models and predictions for the LitCovid dataset. We will use LIT as a general tool to

Table 7.5

Tools and libraries	Description
exBERT https://exbert.net/	Visual Analysis of Transformer Models
ElI5 https://github.com/TeamHG-Memex/eli5	Generic toolbox implementing various black-box, local, and global explanation methods
Alibi https://github.com/SeldonIO/alibi	Implementations of black-box, white-box, local and global explanation methods for classification and regression models.
LIT https://github.com/PAIR-code/lit	The Language Interpretability Tool (LIT) is a visual, interactive model-understanding tool for NLP models.

explore data with KimCNN classifier, understand logistic regression with ELI5 explainers, and transformers-interpret package, which wraps CAPTUM for BERT model explainability.

The Language Interpretability Tool (LIT) is a workbench for diagnosing and improving trained classification and extraction models. LIT is a toolkit and has a modular library—many of the processing and visualization modules can be customized or extended. LIT runs as a server that caches predictions and other data for visualization. The actual visualization modules are written in TypeScript and run in a browser, as shown in Fig. 7.12. A small amount of code must adapt a model to use within LIT, load data, and run the models. LitCovid is an eight-class multi-label dataset with documents drawn from published research studies regarding Covid-19. The eight available classes for each document are as follows: **General**, **Forecasting**, **Transmission**, **Case Report**, **Mechanism**, **Diagnosis**, **Treatment**, and **Prevention**. We only show the predictions for **Forecasting** and **Transmission** to simplify the view for illustration with KimCNN classifier.

Figure 7.12 shows the data table and datapoint editor views on the left side from the Main workspace in LIT. The data table shows available examples with their labels and allows the user to select one or more samples for review or analysis. The datapoint editor window allows the user to modify examples. The Embeddings widget shows a three-dimensional projection of the embeddings for each data sample using UMAP. The current samples selected in the data table get highlighted and accentuated. Selecting a sample or hovering over a sample shows the text (or label) for that point. The projected points get color assigned according to their predicted labels (eight different colors). Individual label predictions or error cases could also be selected. The Group workspace in LIT has many options for working with groups of samples. The default views include scores for the whole dataset and selected samples and a configurable confusion matrix to look at different criteria over different subsets of the data, as shown in Fig. 7.12 at the bottom.

Other group functions in the UI allow for the analysis of individual samples for diagnosis. For example, consider one sample labeled as **Forecasting**, but the model prediction is **General**. We can use the predictions tab to view the distribution

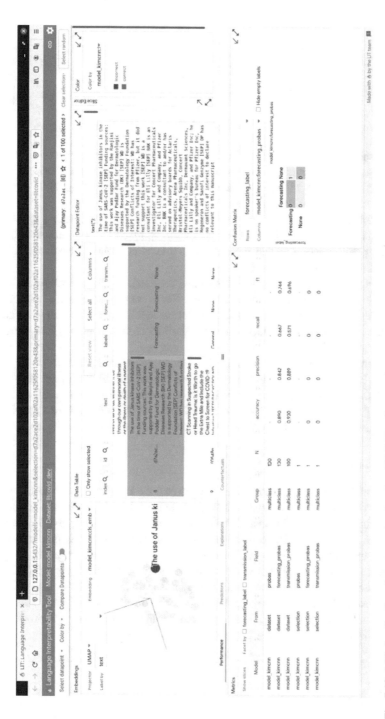

Fig. 7.12 LIT visualization of summary showing data (UMAP), editor, confusion matrix and model outputs

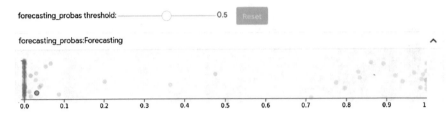

Fig. 7.13 Visualization of model thresholds for individual example and its relative score

of confidence scores for **Forecasting** predictions over the dataset. As with the embedding view, the predictions view dims the samples that are not selected, so we can see that the selected sample scores very low for **Forecasting**, and is rather predicted as **General** as shown in Fig. 7.13.

The Explanations tab provides the ability to use LIME interpretation to assess the weights assigned to individual tokens in the sample. The blue-tinted tokens contribute positive weights, and the red-tinted tokens contribute negative weights as shown in Fig. 7.13.

ELI5 allows to visualize the weights (linear models), the trees, and predictions across wide range of models from scikit-learn, Keras, XGBoost, etc. in a consistent and unified way. The logistic regression model explanation by visualizing weights for features contributing for each class using ELI5 is shown in Fig. 7.14 on the LitCovid dataset.

The transformers-interpret package is a wrapper around Captum, a library developed by the PyTorch team which implements several different explainability techniques. As a result, its visualization methods are a little easier to customize—particularly in displaying multiple records together. In Fig. 7.16 we show LayerGradientXActivation explainability technique on the BERT model for examples.

Observations:

- In Fig. 7.15 it appears that the excerpt is mainly about the sponsorship of the work and does not have sufficient content to determine that the article pertains to the label **Forecasting** and hence misclassified. Figure 7.13 also shows why the confidence of the model for the example is low in prediction. Such explanations help in diagnosing and correcting the data or the model.
- Figure 7.14 shows weights with green-tint and red-tint for how they impact the scoring in positive and negative way, respectively, for each class. It is interesting to see the label **Prevention** having features such as *us*, *many*, *said*, *says*, etc., as the top features indicating the need to better preprocess and clean the dataset using stopwords and lemmatization.

- Figure 7.16 shows similar visualizations on BERT model with Captum explaining the predictions with green-tint and red-tint for positive and negative contributions. The attribution scores and the highlights are interesting from both, explanation and diagnosis, perspectives. For example, the first example with high attribution score is correctly mapped to the class **Case Report** with a score of 0.97 and features such as *transmission, numerical, reports, method*, etc. contribute positively. This gives a good explanation of why the model predicted a certain class. Similarly, the last example though correctly classified as **Treatment** with a low score of 0.41, clearly lacks many positive features that can attribute for its prediction helps in diagnosis.

7.3 Computer Vision

Most state-of-the-art models in various computer vision tasks such as image recognition, object detection, semantic segmentation, instance segmentation, etc. use deep learning architectures. There has been a constant push in the direction of explainability and interpretability in the last several years, especially for computer vision with deep learning. There is a special track in CVPR with experts sharing information and tutorials and presenting exciting research—https://interpretablevision.github.io/. This section will summarize some of the research and explainability techniques, but since it is a vast domain, we shall not cover everything and instead, refer to an excellent book as a resource on the topic [Sam+19]. We will categorize the explainability topics based on the taxonomy presented by Andrea Vedaldi and Ruth Fong based on answering the questions below:

- How to understand the relationship between the input and the output?
- What and how do the deep learning architectures learn?
- How can we improve the deep learning architectures to make them transparent?

For most of the discussion in the topic, our notation will be: The entire input image, represented as $\mathbf{x} \in \mathbb{R}^m$ and the output of the final layer (or any intermediate layer), denoted as $\mathbf{y} \in \mathbb{R}^n$, then the Φ can be the mapping function that corresponds to the entire network, such that:

$$\min_{\mathbf{x}} ||\Phi(\mathbf{x}) - \Phi(\mathbf{x_0})||^2 \tag{7.39}$$

y=Case Report top features		y=Diagnosis top features		y=Forecasting top features		y=General top features		y=Mechanism top features		y=Prevention top features		y=Transmission top features		y=Treatment top features	
Weight?	Feature	Weight?	Feature	Weight?	Feature	Weight?	Feature	Weight?	Feature	Weight?	Feature	Weight?	Feature	Weight?	Feature
+6.211	transmission	+5.001	case	+3.428	treatment	+3.613	patients	+3.944	pandemic	+1.927	says	+1.666	sequences	+1.897	forecasts
+1.558	asymptomatic	+3.982	patient	+2.870	drug	+2.887	diagnosis	+2.721	health	+1.399	people	+1.629	sequencing	+1.721	model
+1.327	cases	+3.790	report	+2.811	vaccine	+2.737	detection	+2.312	care	+1.311	research	+1.602	sars	+1.701	forecasting
+1.216	wuhan	+2.712	old	+2.683	drugs	+2.223	pcr	+2.206	social	+1.086	said	+1.594	genome	+1.304	antis
+1.114	(95	+2.403	day	+2.486	therapeutic	+2.042	clinical	+2.035	surgical	+1.047	elsevier	+1.589	evolutionary	+1.252	models
+1.090	household	+2.144	year	+2.219	trials	+2.030	diagnostic	+2.028	practice	+0.961	health	+1.472	cov	+1.083	number
+1.080	symptomatic	+1.540	infection	+2.172	anti	+1.908	tests	+2.019	lockdown	+0.955	many	+1.337	genomes	+1.064	cases
+1.050	droplets	+1.420	positive	+1.952	antiviral					+0.939	us			+0.967	provinces
... 1536 more positive ...		... 1392 more positive ...		... 1898 more positive ...		... 1944 more positive ...		... 2401 more positive ...		... 1691 more positive ...		... 1712 more positive ...		... 1096 more positive ...	
... 3455 more negative ...		... 3599 more negative ...		... 3093 more negative ...		... 3047 more negative ...		... 2590 more negative ...		... 3300 more negative ...		... 3279 more negative ...		... 3895 more negative ...	
-1.742	patients	-1.694	patients	-2.008	case	-1.959	case	-2.009	sars	-1.270	patients	-1.297	treatment	-1.127	patients
-3.722	<BIAS>	-2.634	<BIAS>	-2.218	health	-1.983	patient	-2.079	severe	-3.939	<BIAS>	-1.428	covid		
						-2.031	pandemic	-2.116	cov			-3.232	<BIAS>	-4.526	<BIAS>

Fig. 7.14 Logistic regression weights visualized using ELI5 on the LitCovid dataset

Fig. 7.15 Visualization of features in text that contribute to the classification. Forecasting example is misclassified as there are few features in the excerpt that contribute to the label

7.3.1   Generating Iconic Examples

One way to examine this is by understanding how much of **x** can be reconstructed by **y**? Mapping input to outputs through phi is a many to one problem as many input data **x** will map to the same **y** or the labels. Reverse mapping the outputs to a set of inputs, known as pre-images, such that they form an equivalence class with the inputs given the network [MV15].

Sampling the input space, starting from a random initialization, and finding the **x** using stochastic gradient descent that minimizes the objective can be one easy technique. Since the network is trained on natural images, starting from random initialization can lead to images that map to the same y but having no characteristics of the natural image that it is trying to reconstruct. Constraining the input to be only in the space of natural images or pseudo-natural images makes the inversion much closer to the desired inputs. Mahendran and Vedaldi approached the problem of inverse representation by adding a "total variation" (TV-norm) regularizer [MV15]. It is given by

$$\min_{\mathbf{x}} ||\Phi(\mathbf{x}) - \Phi(\mathbf{x_0})||^2 + \mathcal{R}(\mathbf{x}) \tag{7.40}$$

Nguyen et al. used an approach to model the distribution and using posterior probability to generate samples [Ngu+17]. This can be represented as:

$$p(\mathbf{x}|\mathbf{y}) = \delta(\Phi(\mathbf{x}) - \mathbf{y}) \cdot p(\mathbf{x}) \tag{7.41}$$

Ulyanov et al. employ constrained optimization to search the pseudo-natural image space for generation [UVL18]. Thus the equation becomes

$$\min_{\mathbf{x}\in\mathcal{R}_{pm}} ||\Phi(\mathbf{x}) - \Phi(\mathbf{x_0})||^2 \tag{7.42}$$

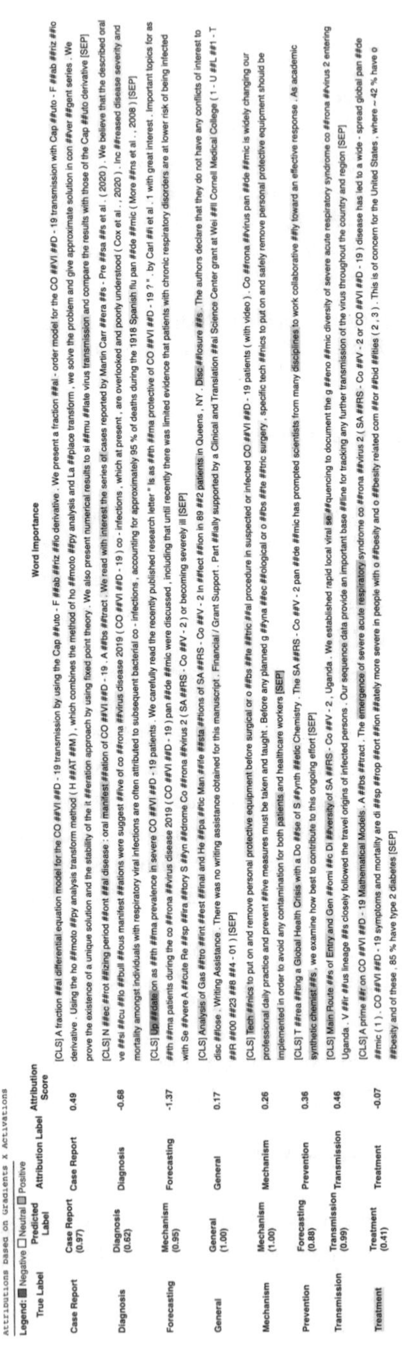

Attributions based on Gradients × Activations

Legend: ■ Negative □ Neutral ■ Positive

True Label	Predicted Label	Attribution Label	Attribution Score	Word Importance
Case Report	Case Report (0.97)	Case Report	0.49	[CLS] ... [SEP]
Diagnosis	Diagnosis (0.62)	Diagnosis	-0.68	[CLS] ... [SEP]
Forecasting	Mechanism (0.95)	Forecasting	-1.37	[CLS] ... [SEP]
General	General (1.00)	General	0.17	[CLS] ... [SEP]
Mechanism	Mechanism (1.00)	Mechanism	0.26	[CLS] ... [SEP]
Prevention	Forecasting (0.88)	Prevention	0.36	[CLS] ... [SEP]
Transmission	Transmission (0.99)	Transmission	0.46	[CLS] ... [SEP]
Treatment	Treatment (0.41)	Treatment	-0.07	[CLS] ... [SEP]

Fig. 7.16 Captum visualization of samples across different labels with highlights on features contributing to the labels using the LayerGradientXActivation technique

Activation Maximization is a technique to visualize inputs to the deep learning networks by maximizing the activation of specific neurons [Erh+09, ZF14a, SVZ14a, MOT15]. For classification, it is iteratively finding parts of the input that the neuron (say in the final output layer) associates with a particular class. This can be written as:

$$\mathbf{x}^* = \arg\min_{\mathbf{x}}(l(\Phi(\mathbf{x}), \Phi_0) + R_\theta(\mathbf{x})) \qquad (7.43)$$

where $l(\Phi(\mathbf{x}), \Phi_0)$ is the loss function between the input $\Phi(\mathbf{x})$ and the target Φ_0 (which can be the weights of features of a particular layer or the final vector of the target class).

There has also been research in building a conditional deep generator network Ψ that takes input $\mathbf{y}$ and recreates an input $\mathbf{x}$ in a fast feed-forward way, minimizing the reconstruction error over large dataset such as ImageNet. That can be written as:

$$\min_{\Psi} \frac{1}{N} \sum_{i=1}^{N} ||\Psi(\Phi(\mathbf{x}_i)) - \mathbf{x}_i||^2 \qquad (7.44)$$

Some variations of this where the goal is to improve the quality of the reconstruction, i.e., a very good prior, replacing the L2 loss with $\mathbf{x}_0 = \mathbf{x}$ (perception loss), $\Phi(\mathbf{x}_0) = \Phi(\mathbf{x})$ (inversion loss) and $p(\mathbf{x}_0) = p(\mathbf{x})$ (GAN loss) [DB15, DB16, Ngu+16, Ngu+17].

7.3.2   Attribution

These techniques try to find essential features or the salient attributes of the image responsible for the output. One of the earliest and the simplest technique is to do sensitivity analysis of the target output neuron (class) to the input pixels by backward propagation and visualizing them as heatmaps [SVZ14b]. Other variants of gradient backpropagation include deconvolution and guided backpropagation. It has been shown that essentially all the techniques are the same except for minor difference on how ReLU is reversed [SVZ14a, ZF14a, Spr+15a]. It was observed that these saliency-based techniques lack channel specificity. A slight modification such as GRAD-CAM, where the backpropagation stops much earlier in the network (typically the first fully connected layer), leads to better channel specificity [Sel+17].

Layer-Wise Relevance Propagation (LRP) and Excitation Backprop are also variants of backpropagation by specifying various rules that define how convolution layers' activations and weights can be combined with the back-propagated signal [Bac+15b, Zha+18]. DeepLIFT approaches the explanation by framing it as the question of importance in terms of differences from a "reference" state. The reference state depends on the task. For example, the reference state for the input can be the presence/absence of a property like a specific object in the image, and the output corresponding to that would be the reference output [SGK17]. Integrated gradients are generalizations of DeepLIFT where the original gradient is used for the non-linearities instead of the average gradient value, preserving the chain rule's validity and thus the implementation invariance [STY17a].

Perturbation-based approaches, in which the inputs are modified and the impacts on outputs are measured, have been considered more principled in interpreting the network than the gradient-based methods. Zeller and Fergus used an occlusion approach where a fixed-size, gray-color occluding square was slid over the input image and the corresponding changes in the feature activation magnitude and/or classification score was observed [ZF14a]. RISE is an extension of the base occlusion where multiple random binary masks slide over the input, and the impact of combining these linearly on the target class/outputs is visualized [PDS18b]. Fong and Vedaldi introduced a meaningful perturbations approach where a minimal mask that perturbs the input to maximize output is learned through optimization [FV19]. Fong et al. extended the technique in their recent work where they compute extremal perturbations which includes a new area constraint along with a parametric family of smooth perturbations, simplifying the optimization problem [FPV19].

LIME is another local explainability method where the image is divided into segments, and then the dataset is sampled with random perturbation like filling with a gray color. The prediction value determines the target of the sample for the accordingly altered input. A weighted regression model is learned, and the weight values on each segment give their importance [RSG16a]. Kernel SHAP combines the concepts of local linear approximations of LIME and Shapley values to explain predictions in superpixel segmented image [LL17a].

7.3.3 Semantic Identification

One general technique is to study the filter-concept overlap, i.e., how the filters and semantic concepts are related through relationships based on the activated patches on the input. Net2Vec learns the concept vectors that describe how a concept is encoded across multiple channels. This method works by probing the network with a concept dataset and learning to perform new tasks using channel activations at a given layer [FV18]. Net2Vec also allows vector arithmetic being performed on the learned concept weights where each dimension is aligned to a filter.

Table 7.6 Tools and Libraries for XAI in CV

Tools and libraries	Description
AIX360 https://aix360.mybluemix.net/	The AI Explainability 360 Python package includes a comprehensive set of algorithms for many domains including CV.
ToolTorch https://github.com/ TooTouch/tootorch	Implementation XAI in Computer Vision (PyTorch) for various attribution, ensemble, and attention methods.
Alibi https://github.com/SeldonIO/alibi	The library provides implementations of various black-box, white-box, local and global explanation methods for classification and regression models.
Captum AI https://captum.ai/	PyTorch based and supports interpretability of models across modalities including vision, text, audio, etc.

7.3.4 Understanding the Networks

To understand what and how the deep networks are learning, one can inspect, analyze, and visualize the networks. Exploring the network can be per layer, per channel, per filter, a combination of single or multiple neurons, etc. Olah et al. in their work introduced methods to understand networks by looking at them as 3d tensors at the individual neuron level, grouped neuron level, from spatial activation level and channel activation level [Ola+18]. The work demonstrated that pairing neuron activation with visualization of that neuron, sorted by the magnitude, brought interpretability to the deep network's hidden layers.

Another way to analyze the role of layers is to use them in transfer learning as feature generators combined with classifiers for different domains and tasks. Razavian et al. use the output of internal layers from the deep learning network with SVM for various tasks such as classification of scenes, attribute detection, and object localization [Sha+14]. Bau et al. introduced "Network Dissection" for quantifying interpretability by estimating the alignment between individual hidden units and a collection of semantic concepts [Bau+17].

7.3.5 Tools and Libraries

Table 7.6 provides important list of tools and libraries that provide implementation for most of the techniques discussed in the sections above.

7.3.6   Case Study

We will explore different explainability techniques using the Fashion MNIST dataset described in Chap. 1. We build a simple convolutional neural network with two convolutional layers as our model. This CNN model reaches the validation accuracy of approx. 92% in 10 epochs.

We use different global and local techniques with the alibi package to understand the reasons for correct classification and fail state analysis. We will use the *Alibi* package for implementing various XAI techniques.

We can quantify the linearity of each layer in a model using the L measure. Higher the L measure, more is the non-linearity present for a given layer. Similarly, we can aggregate and quantify the non-linearity for each class separately. L measure allows us to understand the degree of complexity required to capture each class adequately (Fig. 7.17).

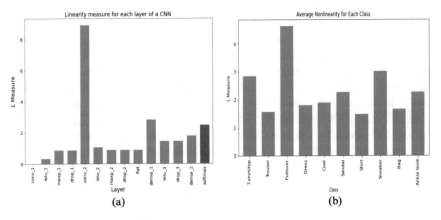

Fig. 7.17 Non-linearity per layer and per class. (**a**) L measure per layer. (**b**) Non-linearity per class

Integrated gradients can be used to understand important features that help in classifying certain class of images and in diagnosing issues for incorrect predictions. Figure 7.18 shows integrated gradient methods showing positive and negative attributions for sample images. Figure 7.19 shows attributions for incorrect predictions.

The counterfactual method finds the minimal amount of distortion required to make the model arbitrarily confident in a different prediction, helping us understand the model's representation of the problem. Figure 7.20 shows how the model prediction changes with small distortion to the original trouser image.

Anchoring is a method that, similarly to integrated gradients, highlights the most informative features in an image. We need to define a few things to prepare for image anchoring. Figure 7.21 shows superpixels and anchors for both correct and incorrect predictions.

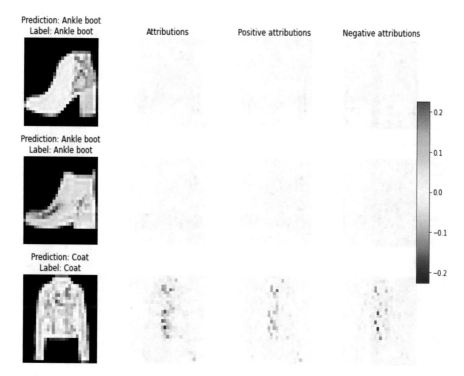

Fig. 7.18 Integrated Gradient explanation for correct predictions

The contrastive explanation method (CEM) attempts to explain the minimally necessary features that influence a prediction. It does this in two ways: pertinent positives and pertinent negatives—the former highlights the pixels that must be present to provide the specified prediction, and the latter highlights those that must be omitted. An autoencoder model is used to generate examples for analyzing CEM techniques. Figure 7.22 shows the original pullover image and the pertinent negative prediction for shirt.

Figure 7.23 shows counterfactuals by prototypes method where the nearest prototype class is used to guide the counterfactual search.

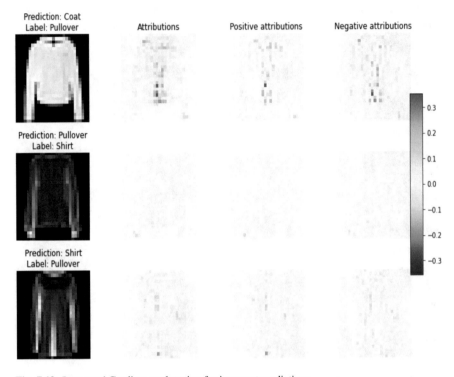

Fig. 7.19 Integrated Gradient explanation for incorrect predictions

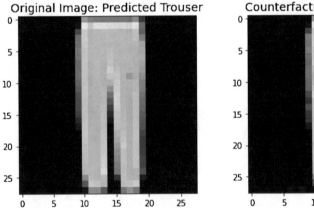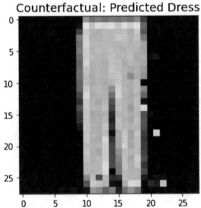

Fig. 7.20 Counterfactual explanations

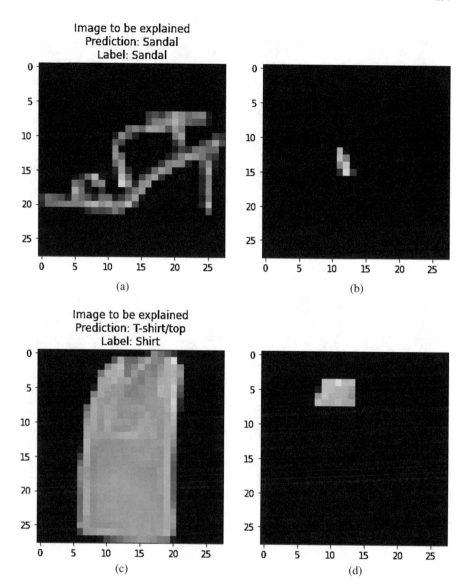

Fig. 7.21 Local explanations using anchors for correct prediction of sandals and incorrect prediction of T-shirt. (**a**) Sandals super pixels. (**b**) Sandals anchors. (**c**) T-shirt super pixels. (**d**) T-shirt anchors

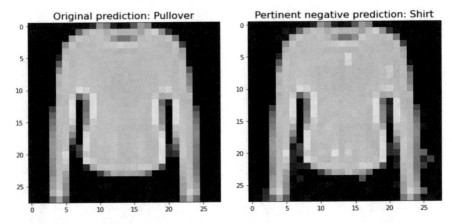

Fig. 7.22 Contrastive explanation

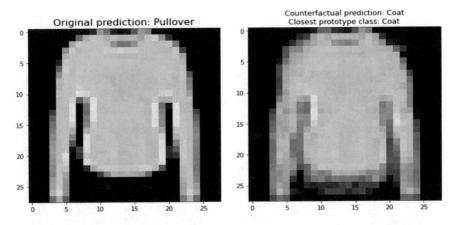

Fig. 7.23 Counterfactuals by prototypes

Observations:

- The L measure plots for different layers in Fig. 7.17 show how the second convolution layer and the first dense layer have high non-linearities. Also, non-linearity by class shows the class pullovers followed by t-shirts and sneakers have more in them.
- Figure 7.18 highlights the first coat's hood and chest, the sneaker's tongue and upper, and the second coat's zipper as an important region in the correct predictions.
- Figure 7.19 brings out regions that are strongly positive and that contribute to the incorrect predictions. For the first image, the model keys on the buttons and predicts a coat instead of a shirt. The second image is a sandal, but the heel's intensity seems to influence the model toward predicting the bag. The third image shows that the model focuses on the hem and zipper, but the zipper's negative attributions were not strong enough to move it toward a coat.
- Figure 7.21 highlights that the intensity of the sandal's strap is the anchor point for predicting it correctly and that the shoulder of this garment is the region confusing the model into classifying it as a t-shirt instead of a shirt. Thus the anchors help diagnose the issues like needing more examples of certain classes with particular features or complexity in the model to capture the regions.
- Figure 7.20 shows how we can trick the model into classifying this pair of trousers like a dress simply by adding a diagonal line of pixels toward the bottom of the garment.
- Figure 7.22 highlights the CEM explanation through the pertinent negative features. We can see that slight changes in pixel intensity around the sleeve could change the prediction from pullover to shirt.
- Figure 7.23 shows how the distortion/blur, when applied to the outline of the shirt, changes the prediction from a pullover to a coat.

References

[Abu+17] A. Abujabal et al., Quint: interpretable question answering over knowledge bases, in *Proceedings of the 2017 Conference on Empirical Methods in Natural Language Processing: System Demonstrations* (2017), pp. 61–66

[AB18] A. Adadi, M. Berrada, Peeking inside the black-box: a survey on explainable artificial intelligence (XAI). IEEE Access **6**, 52138–52160 (2018)

[Bac+15b] S. Bach et al., On pixel-wise explanations for non-linear classifier decisions by layerwise relevance propagation. PLoS ONE **10**, 1–46 (2015). https://doi.org/10.1371/journal.pone.0130140

[BCB14] D. Bahdanau, K. Cho, Y. Bengio, Neural machine translation by jointly learning to align and translate (2014). Preprint, arXiv:1409.0473

[Bau+17] D. Bau et al., Network dissection: quantifying interpretability of deep visual representations, in *Proceedings of the IEEE Conference on Computer Vision and Pattern Recognition* (2017), pp. 6541–6549

[Box+15] G.E.P. Box et al., *Time Series Analysis: Forecasting and Control* (Wiley, Hoboken, 2015)

[Bro59] R.G. Brown, *Statistical Forecasting for Inventory Control* (McGraw/Hill, New York, 1959)

[CMR18] S. Carton, Q. Mei, P. Resnick, Extractive adversarial networks: high-recall explanations for identifying personal attacks in social media posts (2018). Preprint, arXiv:1809.01499

[CRB19] D. Croce, D. Rossini, R. Basili, Auditing deep learning processes through kernel-based explanatory models, in *Proceedings of the 2019 Conference on Empirical Methods in Natural Language Processing and the 9th International Joint Conference on Natural Language Processing (EMNLP-IJCNLP)* (2019), pp. 4028–4037

[Dan+20] M. Danilevsky et al., A survey of the state of explainable AI for natural language processing (2020). Preprint, arXiv:2010.00711

[Don+19] Y. Dong et al., EditNTS: a neural programmer-interpreter model for sentence simplification through explicit editing (2019). Preprint, arXiv:1906.08104

[DB15] A. Dosovitskiy, T. Brox, Inverting convolutional networks with convolutional networks (2015). Preprint, arXiv:1506.02753

[DB16] A. Dosovitskiy, T. Brox, Generating images with perceptual similarity metrics based on deep networks, in *Advances in Neural Information Processing Systems*, vol. 29, ed. by D. Lee et al. (Curran Associates, Red Hook, 2016)

[Dud76] S.A. Dudani, The distance-weighted k-nearest-neighbor rule. IEEE Trans. Syst. Man Cybern. **4**, 325–327 (1976)

[Eba+19] A. Ebaid et al., Explainer: entity resolution explanations, in *2019 IEEE 35th International Conference on Data Engineering (ICDE)* (IEEE, Piscataway, 2019), pp. 2000–2003

[Erh+09] D. Erhan et al., Visualizing higher-layer features of a deep network. Univ. Montreal **1341**(3), 1 (2009)

[FPV19] R. Fong, M. Patrick, A. Vedaldi, Understanding deep networks via extremal perturbations and smooth masks, in *Proceedings of the IEEE/CVF International Conference on Computer Vision (ICCV)*, Oct 2019

[FV18] R. Fong, A. Vedaldi, Net2vec: quantifying and explaining how concepts are encoded by filters in deep neural networks, in *Proceedings of the IEEE Conference on Computer Vision and Pattern Recognition* (2018), pp. 8730–8738

[FV19] R. Fong, A. Vedaldi, Explanations for attributing deep neural network predictions, in *Explainable AI: Interpreting Explaining and Visualizing Deep Learning* (Springer, Berlin, 2019), pp. 149–167

[God+18] F. Godin et al., Explaining character-aware neural networks for word-level prediction: do they discover linguistic rules? (2018). Preprint, arXiv:1808.09551

[Hol04] C.C. Holt, Forecasting seasonals and trends by exponentially weighted moving averages. Int. J. Forecast. **20**(1), 5–10 (2004)

[HA18a] R.J. Hyndman, G. Athanasopoulos, *Forecasting: Principles and Practice* (OTexts, Melbourne, 2018)

[Jia+19] Y. Jiang et al., Explore, propose, and assemble: an interpretable model for multi-hop reading comprehension (2019). Preprint, arXiv:1906.05210

[KLW19b] U. Kamath, J. Liu, J. Whitaker, *Deep Learning for NLP and Speech Recognition*, vol. 84 (Springer, Berlin, 2019)

[LWM19] Q. Li, B. Wang, M. Melucci, CNM: an inter pretable complex-valued network for matching (2019). Preprint, arXiv:1904.05298

[Liu+18] N. Liu et al., On interpretation of network embedding via taxonomy induction, in *Proceedings of the 24th ACM SIGKDD International Conference on Knowledge Discovery & Data Mining* (2018), pp. 1812–1820

[LL17a] S.M. Lundberg, S.-I. Lee, A unified approach to interpreting model predictions, in *Advances in Neural Information Processing Systems*, vol. 30, ed. by I. Guyon et al. (Curran Associates, Red Hook, 2017), pp. 4765–4774. http://papers.nips.cc/paper/7062-a-unified-approach-to-interpreting-model-predictions.pdf

[MV15] A. Mahendran, A. Vedaldi, Understanding deep image representations by inverting them, in *Proceedings of the IEEE Conference on Computer Vision and Pattern Recognition* (2015), pp. 5188–5196

[MSA18] S. Makridakis, E. Spiliotis, V. Assimakopoulos, Statistical and machine learning forecasting methods: concerns and ways forward. PLoS ONE **13**(3), e0194889 (2018)

[Mak+82] S. Makridakis et al., The accuracy of extrapolation (time series) methods: results of a forecasting competition. J. Forecast. **1**(2), 111–153 (1982)

[Mol19] C. Molnar, Interpretable machine learning a guide for making black box models explainable (2019). https://christophm.github.io/interpretablemlbook

[MOT15] A. Mordvintsev, C. Olah, M. Tyka, Inceptionism: going deeper into neural networks (2015)

[Mul+18] J. Mullenbach et al., Explainable prediction of medical codes from clinical text (2018). Preprint, arXiv:1802.05695

[Ngu+16] A. Nguyen et al., Synthesizing the preferred inputs for neurons in neural networks via deep generator networks, in *Advances in Neural Information Processing Systems*, vol. 29, ed. by D. Lee et al. (Curran Associates, Red Hook, 2016)

[Ngu+17] A. Nguyen et al., Plug & play generative networks: conditional iterative generation of images in latent space, in *Proceedings of the IEEE Conference on Computer Vision and Pattern Recognition* (2017), pp. 4467–4477

[Ola+18] C. Olah et al., The building blocks of interpretability. Distill **3**(3), e10 (2018)

[PDS18b] V. Petsiuk, A. Das, K. Saenko, RISE: randomized input sampling for explanation of black-box models, in *British Machine Vision Conference (BMVC)* (2018). http://bmvc2018.org/contents/papers/1064.pdf

[PTS19] P. Pezeshkpour, Y. Tian, S. Singh, Investigating robustness and interpretability of link prediction via adversarial modifications (2019). Preprint, arXiv:1905.00563

[Raj+19] N.F. Rajani et al., Explain yourself! leveraging language models for commonsense reasoning (2019). Preprint, arXiv:1906.02361

[Rem20] P. Remy, N-BEATS: neural basis expansion analysis for interpretable time series forecasting (2020). https://github.com/philipperemy/nbeats

[RSG16a] M.T. Ribeiro, S. Singh, C. Guestrin, "Why should I trust you?" Explaining the predictions of any classifier, in *Proceedings of the 22nd ACM SIGKDD International Conference on Knowledge Discovery and Data Mining* (2016), pp. 1135–1144

[Sam+19] W. Samek et al. (eds.), *Explainable AI: Interpreting, Explaining and Visualizing Deep Learning* (Springer, Berlin, 2019)

[Sel+17] R.R. Selvaraju et al., Grad-CAM: visual explanations from deep networks via gradient-based localization, in *Proceedings of the IEEE International Conference on Computer Vision* (2017), pp. 618–626

[Sen+19] P. Sen et al., HEIDL: learning linguistic expressions with deep learning and human-in-the-loop, in *Proceedings of the 57th Annual Meeting of the Association for Computational Linguistics: System Demonstrations* (2019), pp. 135–140

[Sha+14] A.S. Razavian et al., CNN features off-the-shelf: an astounding baseline for recognition, in *Proceedings of the IEEE Conference on Computer Vision and Pattern Recognition Workshops* (2014), pp. 806–813

[SGK17] A. Shrikumar, P. Greenside, A. Kundaje, Learning important features through propagating activation differences, in *International Conference on Machine Learning, PMLR* (2017), pp. 3145–3153

[SVZ14a] K. Simonyan, A. Vedaldi, A. Zisserman, Deep inside convolutional networks: visualising image classification models and saliency maps (2013). Preprint, arXiv:1312.6034

[SVZ14b] K. Simonyan, A. Vedaldi, A. Zisserman, Deep inside convolutional networks: visualising image classification models and saliency maps (2014)

[Spr+15a] J.T. Springenberg et al., Striving for simplicity: the all convolutional net, in *ICLR (Workshop Track)* (2015). http://lmb.informatik.uni-freiburg.de/Publications/2015/DB15a

[STY17a] M. Sundararajan, A. Taly, Q. Yan, Axiomatic attribution for deep networks, in *Proceedings of the 34th International Conference on Machine Learning - Volume 7, 0. ICML'17. Sydney NSW Australia: JMLR.org* (2017), pp. 3319–3328

[SPR19] A. Sydorova, N. Poerner, B. Roth, Interpretable question answering on knowledge bases and text (2019). Preprint, arXiv:1906.10924

[TL18] S.J. Taylor, B. Letham, Forecasting at scale. Am. Stat. **72**(1), 37–45 (2018)

[UVL18] D. Ulyanov, A. Vedaldi, V. Lempitsky, Deep image prior, in *Proceedings of the IEEE Conference on Computer Vision and Pattern Recognition* (2018), pp. 9446–9454

[Vos+15] N. Voskarides et al., Learning to explain entity relationships in knowledge graphs, in *Proceedings of the 53rd Annual Meeting of the Association for Computational Linguistics and the 7th International Joint Conference on Natural Language Processing (Volume 1: Long Papers)* (2015), pp. 564–574

[Win60] P.R. Winters, Forecasting sales by exponentially weighted moving averages. Manag. Sci. **6**(3), 324–342 (1960)

[Xie+17] Q. Xie et al., An interpretable knowledge transfer model for knowledge base completion (2017). Preprint, arXiv:1704.05908

[ZF14a] M.D. Zeiler, R. Fergus, Visualizing and understanding convolutional networks, in *European Conference on Computer Vision* (Springer, Berlin, 2014), pp. 818–833

[Zha+18] J. Zhang et al., Top-down neural attention by excitation backprop. Int. J. Comput. Vis. **126**(10), 1084–1102 (2018)

[ZHZ18] M. Zhou, M. Huang, X. Zhu, An interpretable reasoning network for multi-relation question answering (2018). Preprint, arXiv:1801.04726

Chapter 8
XAI: Challenges and Future

One of the biggest challenges the XAI field faces is formalizing, quantifying, measuring, and comparing different explanation techniques in a unified way. The evaluation of explanations is an interdisciplinary research covering broad areas of human-computer interaction, machine learning, psychology, cognitive science, and visualization, to name a few. This chapter first highlights some of the recent works in research to categorize and analyze the metrics in a common framework. Finally, we give some predictions on the future based on current trajectories, commercial and open-source trends, and innovations in the field.

8.1 XAI: Challenges

The last few years saw a rapid growth in interpretable and explainable machine learning techniques, discussed in depth in previous chapters. Open issues with explainable methods are formalism for the explanations, assessing the quality of explanations and effective methods to measure them. To evaluate the quality of explanations, we need first to define the attributes of explanations and then map different techniques to these properties. In the last couple of years, significant research has been completed to understand the properties and define metrics qualitatively and quantitatively [Zho+21]. This section will discuss the properties of explanations, categories of explanations, the taxonomy of evaluation and how different explanation techniques map to the properties of explainable systems.

8.1.1 Properties of Explanation

Explainable techniques and interpretable models should provide clarity, simplicity, broadness, completeness, and soundness as critical properties [MKR20, Zho+21]. Clarity means the explanation is unambiguous, i.e., it gives a single reason to similar occurrences and examples. Parsimony implies the explanation is in simple and compact form. Broadness indicates the same explanation applies to a broader range of observations. Completeness of an explanation is assessed through the ability to provide adequate information to compute the output for a given input. Finally, soundness is an indicator of how truthful and correct the explanation is.

8.1.2 Categories of Explanation

Based on the work of the Information Commissioner's Office (ICO) and Webb et al., the explanation itself can be further classified into six categories [KK20, Web+20].

1. Rationale explanation. This answers the "why" part of ML in making decisions. The explanation allows the system user to explain why the decision is flawed or reasons to support the decision.
2. Responsibility explanation. This answers "who" is involved in decisions at various steps of the ML process, i.e., from data to modeling. This type of explanation brings in accountability and traceability.
3. Data explanation. Data plays a vital role in decision-making. This type of explanation focuses on what data is used for training, validating, and testing the model(s).
4. Fairness explanation. This explanation type focuses on ensuring there is no bias and discrimination at various ML decision-making processes. This is a critical area for increasing the trust in AI and ML systems.
5. Safety and performance explanation. This explanation type focuses on ensuring maximum reliability, robustness, and accuracy in decision-making steps.
6. Impact explanation. This type of explanation highlights the impact of the ML systems and decisions at various levels, from individuals to society.

Zhou et al. further classify responsibility and fairness as ethical explanations; the impact explanation is based on the use, but the rest of them are directly related to ML explainabilities [Zho+21]. Finally, they map the explanation types to the different explanation categories. We have made our small changes as given below (Table 8.1):

Table 8.1 Explanation types and categories

Explanation types	Explanation categories
Data explanation	Summary statistics and analysis
	Data visualization (univariate and multivariate)
	Feature relevance and selection
	Prototypical algorithms
Safety and performance explanation	Model selection, validation, and visualization techniques
	Restricted neural network architecture
Rationale explanation	Attribution-based
	Provenance-based
	Surrogate model-based
	Declarative induction-based
	Deep learning and neural visualization techniques

8.1.3 Taxonomy of Explanation Evaluation

Doshi-Velez and Kim divided evaluation approaches into three categories and is a widely accepted taxonomy in explainability evaluation [DK18].

- Application-grounded evaluation (experiments with end-users). In this approach, the subject matter experts or the end-users of the ML system assess how well explanations assist them in decision-making.
- Human-grounded evaluation (experiments with lay humans). In contrast to the application-grounded evaluation, a large group of lay humans instead of domain experts assess the explanation quality in human-grounded evaluation. This approach is more cost-effective but can suffer from significant variance based on the size and type of task.
- Functionality-grounded evaluation (proxies based on a formal definition of interpretability). In this type of evaluation, some quantifiable metric acts as a proxy for the explanation evaluation.

Application-grounded and human-grounded evaluations use human experiments to assess the evaluations, base it on factors such as local-global explanations, the severity of incompleteness in data or models or specifications, time spent by users to understand explanations, and users experience level mapped to the explanations.

Evaluation by humans can be subjective or objective metrics-based. Subjective metrics use questionnaires based on tasks and explanations to create a score based on user trust, confidence, and preference. On the other hand, objective metrics may use physiological and behavioral indicators when subjected to the explanations or use task-related metrics such as time and performance to assess the quality of explanations.

Markus et al. categorized the functionality-grounded evaluation metrics into three categories, viz., model-based, attribution-based, and example-based explanations [MKR20]. Interpretable techniques such as decision trees, Bayesian rule lists, optimal decision trees, etc. discussed in Chaps. 3 and 4, fall under the model-based explanations category. Visualization, feature relevance, and post-hoc attribution techniques such as LIME, SHAP, saliency-based, etc. discussed in Chaps. 5 and 6, apply to attribution-based explanations. Counterfactuals or data-driven explanations, discussed in Chaps. 5 and 6, are part of example-based explanations.

Table 8.2 summarizes the mapping between the functionality-grounded evaluation metrics for measuring the quality of three types of explanation methods

Table 8.2 Explanation types, metrics, and mapping to desired explanation properties

Explanation types	Quantitative metrics	Clarity	Broadness	Simplicity	Completeness	Soundness
Model-based explanations	Model size [Gui+18, MCB20b]			Y		
	Runtime operation counts [Sla+19]			Y		
	Interaction strengths [MKR20, MCB20b]			Y		
	Main effect complexity [MCB20b]			Y		
	Level of disagreement [Lak+17]	Y				Y
Attribution-based explanations	Monotonicity [NM20]					Y
	Sensitivity [Yeh+19, STY17b, NM20]					Y
	Effective complexity [NM20]		Y	Y		
	Remove and retrain [Hoo+18]					Y
	Recall of important features [RSG16b]					Y
	Implementation invariance [STY17b]					Y
	Selectivity [MSM18]					Y
	Continuity [MSM18]	Y				
	Sensitivity-n [Anc+17]					Y
	Mutual information [MCB20b]		Y	Y		Y
Example-based explanations	Non-representativeness [MCB20b]			Y	Y	
	Diversity [MCB20b]			Y		

with desired properties sought in explanations given by Zhou et al. with some modifications. We use the letter Y to illustrate that the metric successfully helps in quantifying the respective property.

8.2 Future

The future is AI-driven computing, and there is little doubt that the explainability of the models will play a significant role in shaping it. Based on the current trends in the academic, commercial, and open-source community, we identify some of the possible changes and directions XAI will take in the next few years.

8.2.1 Formalization of Explanation Techniques and Evaluations

The lack of formalism in explanation techniques and customized explanation for different users in different domains has posed severe challenges in evaluating and comparing explanation techniques in AI. Nevertheless, there has been some progress at the domain level or in the particular sub-field of machine learning. For example, Hardt et al. propose a generalized framework for quantifying and reducing discrimination in supervised learning settings for tabular-based data [HPS16]. Islam et al. show how to leverage domain knowledge and infuse this into black-box models for better explainability in finance and cybersecurity domains [Isl+19b, Isl+19a]. However, a generic framework that addresses measuring, quantifying, comprehensibility, etc. will make the techniques and evaluation more formal. Datasets, benchmarks, and competitions will further fuel innovation and standardization.

8.2.2 Adoption of Interpretable Techniques

In her work "Stop Explaining Black-Box Machine Learning Models for High Stakes Decisions and Use Interpretable Models Instead," Rudin makes a case for broader adoption of interpretable machine learning models as opposed to explainable techniques [Rud19d]. The *Rashomon* set argument is: If there is a large set of models with relatively reasonable accuracy, there has to be an interpretable model which is both accurate and interpretable. This argument might be true for most classes of problems in different domains. In Chaps. 3 and 4, we covered most of the techniques in interpretable machine learning, demonstrating its usefulness and some of the challenges on classification and regression problems. Applying interpretable machine learning techniques based on optimality for high-dimensional data such as

NLP can be challenging due to the search space. Nonetheless, if the open-source community, policy-makers, and subject-matter experts all push to embrace the path of interpretable models, there will be more adoption and deployments in various industries.

8.2.3   Human-Machine Collaboration

Most explanations have to be comprehensible and understood by the users who may have a different level of expertise. In addition, depending on their backgrounds, there may be more questions or feedback that the user might provide for assessing the explanations. The experts from the community of Human-Computer Interaction (HCI) need to play an essential role in capturing these user-explanation interactions. In the future, the system has to capture these interactions, automate the process, and gather the learnings to improve the model and explanations rather than the siloed process that exists today [Rab+21].

8.2.4   Collective Intelligence from Multiple Disciplines

Islam et al., in their survey, point out that explanations and interpretations touch different disciplines such as philosophy, psychology, sociology, cognitive science, etc., and the need of the hour is the multidisciplinary research that can further advance XAI [Rab+21].

8.2.5   Responsible AI (RAI)

Responsible AI (RAI) is an emerging field that combines many aspects of explainability, interpretability, trust, ethics, fairness, privacy, and security to overcome the issues facing the current machine learning-based operations [Arr+20b, Sch+20]. Many industries, especially healthcare and finance, are making a conscious move to adopt the RAI framework from policy perspective. Many AI companies like Microsoft, Google, Amazon, Facebook are investing in developing policies, toolkits, and libraries to support RAI.

8.2.6   XAI and Security

Vigano and Magazzeni's position paper on explainable security (XSec) discusses every aspect of how explainable AI can contribute and benefit the security community from users to applications such as threat modeling and preventing exploiting vulnerabilities [VM20b]. As more and more companies leverage machine learning and explainable AI, there is also a threat of attacks on these explainable AI and interpretable machine learning models, as discussed by Kuppa and Le-Khac [Kup+19]. XAI and its impact on security will be critical for the research community, government, and the commercial world in the coming years.

8.2.7 Causality and XAI

One can view interpretable machine learning as a method focusing on the association between data and outcome instead of causality. On the other hand, causal machine learning and inference using many causality-oriented methods focus on causality and play a significant role in interpretable machine learning [Mor+20]. Causal ML is an emerging field from both machine learning and interpretability viewpoint and will play an essential role in the future.

8.3 Closing Remarks

We hope that the readers found the content both informative and helpful. We hope we have enabled the readers to get an understanding of various techniques in XAI with practical examples and case studies using open-source libraries.

References

[Anc+17] M. Ancona et al., Towards better understanding of gradient-based attribution methods for deep neural networks (2017). Preprint, arXiv:1711.06104

[Arr+20b] A.B. Arrieta et al., Explainable Artificial Intelligence (XAI): concepts, taxonomies, opportunities and challenges toward responsible AI. Inf. Fusion **58**, 82–115 (2020)

[DK18] F. Doshi-Velez, B. Kim, Considerations for evaluation and generalization in interpretable machine learning, in *Explainable and Interpretable Models in Computer Vision and Machine Learning* (Springer, Berlin, 2018), pp. 3–17

[Gui+18] R. Guidotti et al., A survey of methods for explaining black box models. ACM Comput. Surv. **51**(5), 1–42 (2018)

[HPS16] M. Hardt, E. Price, N. Srebro, Equality of opportunity in supervised learning (2016). Preprint, arXiv:1610.02413

[Hoo+18] S. Hooker et al., A benchmark for interpretability methods in deep neural networks (2018). Preprint, arXiv:1806.10758

[Isl+19a] S.R. Islam et al., Domain knowledge aided explainable artificial intelligence for intrusion detection and response (2019). Preprint, arXiv:1911.09853

[Isl+19b] S.R. Islam et al., Infusing domain knowledge in ai-based "black box" models for better explainability with application in bankruptcy prediction (2019). Preprint, arXiv:1905.11474

[KK20] E. Kazim, A. Koshiyama, Explaining decisions made with AI: a review of the co-badged guidance by the ICO and the Turing Institute (2020). Available at SSRN 3656269

[Kup+19] A. Kuppa et al., Black box attacks on deep anomaly detectors, in *Proceedings of the 14th International Conference on Availability, Reliability and Security* (2019), pp. 1–10

[Lak+17] H. Lakkaraju et al., Interpretable & explorable approximations of black box models (2017). Preprint, arXiv:1707.01154

[MKR20] A.F. Markus, J.A. Kors, P.R. Rijnbeek, The role of explainability in creating trustworthy artificial intelligence for health care: a comprehensive survey of the terminology, design choices, and evaluation strategies. J. Biomed. Inf. **113**, 103655 (2020)

[MCB20b] C. Molnar, G. Casalicchio, B. Bischl, Interpretable machine learning–a brief history, state-of-the-art and challenges (2020). Preprint, arXiv:2010.09337

[MSM18] G. Montavon, W. Samek, K.-R. Müller, Methods for interpreting and understanding deep neural networks. Digit. Signal Process. **73**, 1–15 (2018)

[Mor+20] R. Moraffah et al., Causal interpretability for machine learning problems, methods and evaluation. ACM SIGKDD Explor. Newslett. **22**(1), 18–33 (2020)

[NM20] A.-p. Nguyen, M.R. Martínez, On quantitative aspects of model interpretability (2020). Preprint, arXiv:2007.07584

[Rab+21] S.R. Islam et al., Explainable artificial intelligence approaches: a survey (2021). e-Prints, arXiv–2101

[RSG16b] M.T. Ribeiro, S. Singh, C. Guestrin, "Why should I trust you?" Explaining the predictions of any classifier, in *Proceedings of the 22nd ACM SIGKDD International Conference on Knowledge Discovery and Data Mining* (2016), pp. 1135–1144

[Rud19d] C. Rudin, Stop explaining black box machine learning models for high stakes decisions and use interpretable models instead. Nat. Mach. Intell. **1**(5), 206–215 (2019)

[Sch+20] D. Schiff et al., Principles to practices for responsible AI: closing the gap (2020). Preprint, arXiv:2006.04707

[Sla+19] D. Slack et al., Assessing the local interpretability of machine learning models (2019). Preprint, arXiv:1902.03501

[STY17b] M. Sundararajan, A. Taly, Q. Yan, Axiomatic attribution for deep networks, in *International Conference on Machine Learning, PMLR* (2017), pp. 3319–3328

[VM20b] L. Vigano, D. Magazzeni, Explainable security, in *2020 IEEE European Symposium on Security and Privacy Workshops (EuroS&PW)* (IEEE, Piscataway, 2020), pp. 293–300

[Web+20] M.E. Webb et al., Machine learning for human learners: opportunities, issues, tensions and threats, in *Educational Technology Research and Development* (2020), pp. 1–22

[Yeh+19] C.-K. Yeh et al., On the (in) fidelity and sensitivity for explanations (2019). Preprint, arXiv:1901.09392

[Zho+21] J. Zhou et al., Evaluating the quality of machine learning explanations: a survey on methods and metrics. Electronics **10**(5), 593 (2021)

Printed in the United States
by Baker & Taylor Publisher Services